# Chasing Aces
## in Middle Grade Math

By Phil Atkinson
Illustrated by Pat Bochnak

CHASING ACES



1210 SW 23rd PL • Ocala, Florida 34471 • Phone 352-622-1825
Website: www.atlantic-pub.com • Email: sales@atlantic-pub.com
SAN Number: 268-1250



Library of Congress Cataloging-in-Publication Data

Names: Atkinson, Philip, author.
Title: Chasing aces in middle-grade math : a guide to teaching through the lens of disc golf / by Philip Atkinson.
Description: Ocala, Florida : Atlantic Publishing Group, Inc., [2020] | Includes bibliographical references and index. | Summary: "This is a middle-grade math teaching reference for PE and math teachers alike. The lessons are thorough and follow Common Core Standards for Oregon, Washington, Idaho, Utah, Colorado, Arizona, and Pennsylvania/Georgia"— Provided by publisher.
Identifiers: LCCN 2020005175 | ISBN 9781620237199 (paperback) | ISBN 9781620237205 (ebook)
Subjects: LCSH: Mathematics—Study and teaching (Middle school)—United States. | Science—Study and teaching (Middle school)—United States. | Disc golf. | Active learning. | AMS: Mathematics education — Education and instruction in mathematics — Teaching methods and classroom techniques. | Mathematics education — Education and instruction in mathematics — Teaching units and draft lessons.
Classification: LCC QA13 .A895 2020 | DDC 372.70973—dc23
LC record available at https://lccn.loc.gov/2020005175



Printed in the United States

PROJECT MANAGER: Kassandra White
INTERIOR LAYOUT: Nicole Sturk

To my dear friend and colleague Corey Carlson who helped with the innovation for this book. His knowledge, skill and love for the sport inspired me to complete this work while enjoying the game. It's our hope many students will excel in math and become competent in the fun and exciting sport of Disc Golf. Thanks Corey!

# When Learning Begins

❖ **What to consider when creating a structured learning environment.**

❖ **When having fun is predictable, students can't wait to be in your class for instruction and you can provide guidance in real math application.**

My supervising teacher and mentor told me as a young student teacher, "If you aren't having a blast, rest assured, neither are they."

Learning math is notorious for creating fear in many children. They want to know and understand why they are doing something and if they will ever need to use it again in real life. Building confidence and a natural propensity for problem solving is a math teacher's objective. We often hear from our students, "My parents were bad in math, and they kind of passed that along to me." Right or wrong, this is a consequence of prior failures in math classrooms where students were not led to arrive at the correct solution by using varying strategies to get there. Experiencing the process can be a creative pathway to success.

Positive learning comes from exposure and practical involvement with real world situations and experiences. It should place an emphasis on hands-on innovative strategies and skills that let students engage in the learning process.

Our goal is to provide novel ways to involve students in experiencing math, science, and technology through the sport of disc golf.

If you are a teacher that loves fresh air, loves the great outdoors, loves to see math in any situation, and appreciates the science that backs it up, then this is just for you. Our easy to follow lesson plans will have you immersed in math, science, P.E., and Disc Golf in no time. That's a win-win situation!

Grab a pencil, a clipboard, some paper, your kicks. and shades, and you are on your way to opening a whole new world for your students. Experience math, science, technology, and disc golf, and enjoy the excitement of the Professional Disc Golf Association (PDGA) right out back.

# Table of Contents


| LESSON #1 | OBJECTIVES |
|---|---|
| Page 1 | *Concepts* |
| Math | Mean, Median, Mode, Range |
| Disc Golf | Throwing Styles |
| Science | Newton's Laws of Motion |
| Enrichment | Integers, Absolute Value,<br>Index Card Number Line Activity |
| Technology | Using the Excel Sort Command |
| Page 2 | *Anticipatory Vocabulary* |
| Math | Mean, Median, Mode, Range |
| Science | Motion, Center of Mass, Speed, Velocity,<br>Acceleration, Inertia, Momentum,<br>Newton's Laws |
| Page 3 | *Anticipatory Vocabulary* |
| Disc Golf | Stability, Over- Stable, Under-Stable,<br>Turn, Fade, Rotation, Spin Rate |
| Page 4 | *Disc Golf Games* |
| Game A | Throwing Race |
| Game B | Disc Flip |
| Page 5 | *Activity* |
| Throw for Distance/Time | Collect Data to Graph<br>Mean, Median, Mode |
| | *Math in Action* |
| F = m x a | Force Needed to Accelerate a Disc |
| Page 6 | *Math in Action* |
| A = f ÷ m | Finding Acceleration Rates |

# Table of Contents

| LESSON #2 | OBJECTIVES |
|---|---|
| Page 7 | *Concepts* |
| Math | Bar Graphs, Line Graphs |
| Disc Golf | Putters, Mid-Range, Drivers |
| Science | $D = v \cdot t$    $V = d/t$    $T = d/v$ |
| Enrichment | Scoring Using Integers |
| Page 8 | *Concepts* |
| Technology | Using Excel to Create Line/Bar Graphs |
| | *Anticipatory Vocabulary* |
| Math | Scales, Legends, Grid Lines |
| Science | Motion, Linear Motion, Speed, Velocity, Acceleration, Newton's Laws |
| Page 9 | *Anticipatory Vocabulary* |
| Disc Golf | Putter/Driver Stability |
| | *Disc Golf Games* |
| Game A | Water Transport |
| Game B | Disc Tag |
| Page 10 | *Activity* |
| Throw for Distance Practice | Pacing to Calculate Distance in Yards |
| Page11 | *Excel Graphing* |
| Instructions for Creating a Vertical Bar Graph in Excel | Showing Averages from Distance Practice Activity |
| Page 12 | *Grade Rubric* |
| Assessment Chart | For Distance Practice Activity Results |
| Page 13 | *Math in Action* |
| $V = d / t$ | Calculating Speed |
| $T = d / v$ | Calculating Length of Time |

# Table of Contents


| LESSON #3 | OBJECTIVES |
|---|---|
| Page 14 | *Concepts* |
| Math | Scatter Plots, Histograms, Percentages, Angle Sizes, Angle Classifications |
| Disc Golf | 1-Step, 2-Step Cross and Throw |
| Science | Gravitation, Terminal Velocity,<br>Escape Velocity, Free Fall |
| Enrichment | Bar Graphs with Constants of Proportionality in Percentages Using Means/Extremes |
| Page 15 | *Concepts* |
| Technology | Using Excel to Create a Scatter Plot or Histogram to Track Distances |
| | *Anticipatory Vocabulary* |
| Math | Frequency Table, Frequency of an Interval, Histogram, Scatter Plot, Acute, Obtuse |
| Science | Newton's Laws of Gravitation, Terminal Velocity, Escape Velocity, Free fall |
| Page 16 | *Anticipatory Vocabulary* |
| Disc Golf Rules | Tee Off, Initiating Position, Maintaining the Lie, Throwing From a Fixed Position, Holing Out, Out of Bounds, Discs for Play, Using Courtesy |
| Page 17 | *Disc Golf Games* |
| Game A | Hula Hoop Disc Throw |
| Game B | Castle Toss |
| Page 18 | *Math in Action* |
| Percentage Throws | Calculating Percentages Using Constants of Proportionality |

## Table of Contents

| LESSON #4 | OBJECTIVES |
|---|---|
| Page 19 | *Concepts* |
| Math | Solving Equations, Stem and Leaf Plots |
| Disc Golf | Using Mid-Range Discs and Drivers (Practice Throws) |
| Science | Levers, Fulcrum, Resistance Arm, Effort Arm, Machines, Efficiency, Mechanical Advantage, Work |
| Enrichment | $MA = L_e / L_r$ Effort Arm ($L_e$) Resistance Arm ($L_r$) |
| Technology | Using Excel to Create a Stem and Leaf Plot |
| Page 20 | *Anticipatory Vocabulary* |
| Math | Scales, Legends, Stem-and-Leaf Plot |
| Science | 1st, 2nd, 3rd Class Levers, Fulcrum, Resistance Arm, Effort Arm, Machines, Efficiency |
| Disc Golf | Disc Information and Features |
| Page 21 | *Disc Golf Games* |
| Game A | Disc Tic Tac Toe |
| Game B | Disc Bowling |
| Page 22 | *Math in Action* |
| $MA = F_r / F_e$ | Finding the Mechanical Advantage of a Crow Bar |
| Stem and Leaf Plots | A Graph Used to Record, List and Keep Tract of Specific Data |

# Table of Contents


| LESSON #5 | OBJECTIVES |
|---|---|
| Page 23 | *Concepts* |
| Math | Circumference, Area |
| Disc Golf | Putting Practice |
| Science | Circular Motion, Centripetal Force, Centrifugal Force, Newton's $2^{nd}$ Law |
| Enrichment | Name and Identify Central Angles, Radius, Diameter, Chords |
| Technology | Create a Pie Graph Using Excel |
| Page 24 | *Anticipatory Vocabulary* |
| Math | Circumference, Area, Radius, Diameter, Central Angle, Chord |
| Science | Centripetal Acceleration, Centripetal/ Centrifugal Forces, Newton's $2^{nd}$ Law |
| Disc Golf | Underhand Putt, Overhand Putt |
| Page 25 | *Disc Golf Games* |
| Game A | Disc Horse |
| Game B | Disc Bocce |
| Page 26 | *Putting Practice Activity* |
| Angle Shots | Students Putt in Pairs from Three Angles- Left 60°, Center 90°, Right 60° (10 ft., 15 ft.) |
| Page 27 | *Excel Graphing* |
| Instructions for Creating a Horizontal Bar Graph in Excel | Showing Averages from Putting Practice Activity |
| Page 28 | *Grade Rubric* |
| Assessment Chart | For Putting Practice Activity results |
| Page 29 | *Math in Action* |
| $A = V^2 / r$ | Finding the Centripetal Acceleration of an Object |

# Table of Contents

| LESSON #6 | OBJECTIVES |
|---|---|
| Page 30 | *Concepts* |
| Math | Rate, Unit Rate, Average Rate |
| Disc Golf | Putting Practice Across Width of a Football Field |
| Science | Lift, Drag, Gravity, Air resistance, Angular Momentum |
| Enrichment | Rates and Unit Rates to Evaluate and Calculate Average Rate as<br>D / V |
| Technology | In Excel Create any Graph to Calculate Percentage of Successful/Unsuccessful Shots<br>(Bar, Line, Scatter, Histogram, Frequency, or Stem-and Leaf) |
| Page 31 | *Anticipatory Vocabulary* |
| Math | Rate, Unit rate, Average Rate |
| Science | Lift, Drag, Gravity, Air resistance, Angular Momentum |
| Disc Golf Diagram of Aerodynamic Components | Compare and Contrast Various Types of Movement Within a Range of Mathematical Components that Describe How a Disc Moves |
| Page 32 | *Disc Golf Games* |
| Game A | Disc Golf Dodge Ball |
| Page 33 | *Math in Action* |
| $R = D / T$ | Calculating a Disc's<br>Rate of Ascent |
| $D = r \cdot t$ | Calculating the Distance Walked |

| LESSON #7 | OBJECTIVES |
|---|---|
| Page 34 | *Concepts* |
| Math | Pythagorean Theorem |
| Disc Golf | Using the Pythagorean<br>Theorem to Find the Exact Placement for Tee-Off Pad |
| Science | Converting Distance from Feet to Meters |
| Enrichment | Create a Ladder for Converting Weight (grams), Capacity (liters), Distance (meters) |
| Technology | In Excel Create a Graph to Show Missing Mileage on Parts of a Road Map Using the Pythagorean Theorem |
| Page 35 | *Anticipatory Vocabulary* |
| Math | Legs, Hypotenuse, Pythagorean Theorem |
| Science | Gram, Liter, Meter |
| Disc Golf Concepts | Targets, Tee Pads, Signage, Topography |
| Page 36 | *Disc Golf Games* |
| Game A | Ultimate Disc Golf |
| Page 37 | *Math in Action* |
| Manipulating Conversions Using Metrics | Conversion Activity Chart for Weight (grams), Capacity (liters), Distance (meters) |

## Table of Contents

| | |
|---|---|
| Page 38 | *Disc Golf Complex Project* |
| Using Formulas for Geometric Shapes | Calculating Cost of Materials for Sod, Pavers, and Mulch |
| Page 39 | *Designing a Disc Golf Complex Courtyard* |
| Using Formulas to Create Areas in a Courtyard | Calculating Cost of Materials for<br>Sod, Pavers, and Mulch |
| Page 40 | *Math in Action* |
| Areas of Triangles and Trapezoids | Using Equations Chart Showing How to Solve for the Areas of Triangles and Trapezoids |
| Page 41 | *Disc Golf Courtyard Worksheet* |
| Creating a Courtyard on Graph Paper with Areas and Dimensions for each Sector | Pavers- Patios, Tee Pads, Walkways, Barbeque Pits, Hot Tubs<br>Bark Mulch- Coverings for<br>Flower Garden, Shrubs, or Surrounding Trees<br>Grass Sod- Ground Cover |
| Page 42 | *Extra Shape worksheet* |
| Extra Shapes not Required in Directions | Worksheet for Extra Shapes |
| Page 43 | *Disc Golf Complex Courtyard Grade Rubric* |


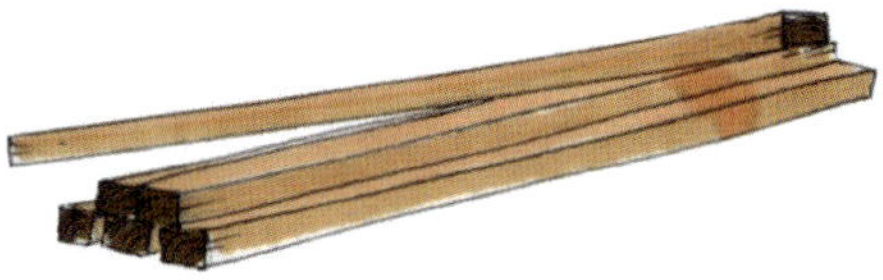

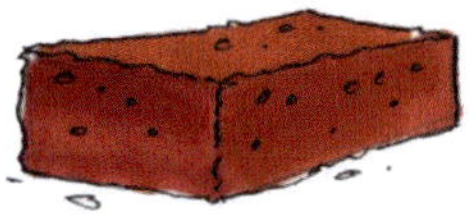

## Using Formulas for Geometric Shapes

| <u>Trapezoids</u> | <u>Rectangles</u> | <u>Squares</u> | <u>Triangles</u> |
|---|---|---|---|
| $\frac{1}{2}h(b_1 + b_2)$ | $l(w)$ or $b(h)$ | $s^2$ | $\frac{1}{2}\,b(h)$ |

## Table of Contents


| LESSON #8 | OBJECTIVES |
|---|---|
| Page 44 | *Concepts* |
| Math | Construct and Diagram a Model of Muscle Groups |
| Disc Golf | Muscle Groups Used During Disc Golf Warm Up Calisthenics |
| Science | Identify Muscle Groups of the Upper, Mid, and Lower Torso Used in Exercising |
| Enrichment | Using a Blood Pressure Cuff to Take a Blood Pressure Reading and a Stethoscope to Measure Heart Rates |
| Technology | Using Excel to Create Line/Bar Graphs and Relative Frequency Tables to Show Blood Pressures and Heart Rates (pre, during, and post) While Exercising |
| Page 45 | *Exercising Muscle Groups* |
| Upper, Mid, Lower Torse | Upward Caterpillar, Downward Caterpillar, Hurdler's Stretch, Butterfly, Spiral Football Stretch, Standing Quad Stretch, Lateral Shoulder Stretch, Back Scratchers, Lunges |
| Page 46 | *Frontal Muscle Anatomy Diagram* |
| Page 47 | *Warm Up Calisthenics* |
| Exercise Explanations | Upward Caterpillar, Downward Caterpillar, Hurdler's Stretch, Butterfly |
| Page 48 | Spiral Football Stretch, Standing Quad Stretch, Lateral Shoulder Stretch, Back Scratchers, Lunges |
| Page 50-51 | *Disc Grips* |
| Finger Positioning | Three-Finger Grip, Power Grip, Fan Grip, Split Finger, Stack Grip, Thumb Positioning |

# Table of Contents

| LESSON #9 | OBJECTIVES |
|---|---|
| Page 52 | *Concepts* |
| Math | Tree Diagram Showing Sample Space for a Healthy Meal |
| Disc Golf | Identifying Food Groups at Each Hole of Play to Provide for a Healthy Meal of Appropriate Percentages of Fats, Carbs, and Proteins on the Nutrition Wheel |
| Science | Create a Sample Space of Foods that Meet the FDA Standards for a Balanced Daily Diet of 2000 Calories |
| Enrichment | Creating a Weekly Plan for a Healthy 2000 Calorie Diet Placed on a Visual Pie Graph |
| Technology | Students Will Create a Regression Analysis Chart in Excel to Show Variations in Food Groups to Maximize Fruits and Vegetables for a Healthy Diet Plan |
| Page 53 | *Anticipatory Vocabulary* |
| Math | Populations, Sample Space,<br>Tree Diagrams |
| Science | Fats, Carbohydrates, Proteins |
| Disc Golf Concepts | Joules, Law of Conservation of Energy/ Momentum, Mechanical Advantage, Mechanical Energy, Resistance Force |
| Page 54 | *Food Groups* |
| Foods and Combinations of When to Eat Cooked v.s. Raw | Fresh Fruits, Starches,<br>Animal Protein, Nuts, Seeds, Dried Fruits, Raw and<br>Non-Starchy Vegetables |
| Page 55 | *Math in Action* |
| Food Diagram for Percentages of Carbs, Fats, and Proteins | Choosing Healthy<br>Eating Habits and Categorizing Food As:<br>Best, Better, Good, Fair, Poor |

## Table of Contents


| LESSON #10 | OBJECTIVES | |
|---|---|---|
| Page 56 | *Concepts* | |
| Math | Dimensional Analysis | |
| Disc Golf | Doubling and Tripling a No Bake Cookie Recipe<br>(Disc Shaped Cookies) | |
| Science | Converting Ounces, Cups, Pints,<br>Quarts, Gallons | |
| Enrichment | Using Variations to Convert SI Standard Units to Metric Units | |
| Technology | Create a Power Point Presentation Showing How to Convert from SI Standard Units to Metric Units | |
| Page 57 | *Anticipatory Vocabulary/Formulas* | |
| Math | Converting Celsius<br>$C = 5/9(F - 32)$ | Converting Fahrenheit<br>$F = 1.8 \times C + 32$ |
| Science | Metrics to SI<br>in x 2.54 = cm | Miles to Kilometers<br>mi x 3 ÷ by 2 = km |
| Disc Golf Concepts | Kilometers to Miles<br>k x 0.6214 = mi | Ounces to Grams<br>oz x 28.35 = g |
| Page 58 | *Math in Action* | |
| Disc Golf No Bake Cookies | Create a Triple Batch of No Bake Disc Shaped Cookies by Multiplying Mix Numbers | |

## Table of Contents

| LESSON #11 | OBJECTIVES | |
|---|---|---|
| Page 59 | *Concepts* | |
| Math | Exponential Notation | |
| Disc Golf | Keeping Track of Local Weather Conditions for a Week to Ensure Safe Play | |
| Science | Thermometers, Barometers, and Anemometers | Used to Check Wind Speed and Wind Chill |
| Enrichment | Identifying Weather Symbols | Using the International Weather Chart Online |
| Technology | Create a Forecast Model for a Major City from a Weather App | |
| Page 60 | *Anticipatory Vocabulary* | |
| Math | Power, Base, Exponents | |
| Science | Thermometer, Barometer, Anemometer | Sling Psychrometer,<br>Wind Vane, Rain Gauge, Hygrometer,<br>Weather Satellites |
| Disc Golf Concepts | Head, Tail and Cross Winds | |
| Page 61 | *Weather Station Symbols Chart* | |
| Page 62 | *Math in Action* | |
| Reading a Weather Forecast Model<br>Determining and Evaluating Weather Conditions | Temperature, Present Weather,<br>Dew Point, Low Cloud Types,<br>Pressure Change | Pressure Tendency,<br>Wind Speed/Direction,<br>Barometric Pressure,<br>High Cloud Types,<br>Cloud Coverage |
| Page 63 | *Wind Chill Index Chart* | |
| Calculating Wind Speed and Temperature to Find the Wind Chill | Use the Wind Chill Formula to Find the Actual Temperature | |
| Calculating Wind Speed and Temperature to Find the Wind Chill | Use the Wind Chill Formula to Find the Actual Temperature<br>Wind Chill (°F) =<br>$35.74 + 0.6215T - 35.75 (V^{0.16}) + 0.4275T (V^{0.16})$ | |


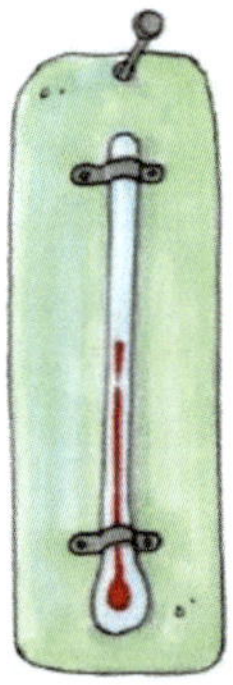

# Table of Contents

| LESSON #12 | OBJECTIVES | |
|---|---|---|
| Page 64 | *Concepts* | |
| Math | Parallels of<br>Latitude and Longitude | |
| Disc Golf | Find Disc Golf Tournaments Played Internationally<br>Use Latitude and Longitude to Find Exact Locations of Countries on a Political Map | |
| Science | Determine Degrees, Minutes, and Seconds to Establish Exact Locations<br>$1° = 60' = 3600''$ | For Finer Accuracy Use Sexagesimal Notation Where Fractions of Seconds are Given by a Decimal Point |
| Enrichment | Given Several Latitude and Longitude Coordinates with Their Minutes, Degrees, and Seconds, Find Tournaments on a Map | |
| Technology | Use the Following iPhone Apps to Practice Locating Specific Landmarks from a Given Position | [GPS Kit] [MotionX GPS] [Gaia GPS] [iTopo Maps] [Trails – GPS Tracker]<br>[Sun Compass] |
| | Practice Navigating to a Location of Your Choice Successfully | [Park Maps] [Accu Terra] [Geocaching]<br>[Elevation Pro] |
| Page 65 | *Anticipatory Vocabulary* | |
| Math | Coordinate Plane, Ordered Pairs | |
| Science | Degrees, Minutes, Seconds, Hyzer Angle, Anhyzer Angle | |
| Disc Golf Concepts | Over Stable | Spike/Straight Hyzer<br>Anhyzer Flex |
| | Stable | Hyzer Flip<br>Distance Flex<br>Max Distance |
| | Under Stable | Hyzer Flip<br>Anhyzer<br>Roller |
| Page 66 | *Find Tournaments on a Political Map* | |
| Sexagesimal Formulas | Convert to a Decimal | Convert to D°M'S" |
| Find Locations | Identify City Name | Identify Country |

# Table of Contents

| LESSON #13 | OBJECTIVES | |
|---|---|---|
| Page 67 | *Concepts* | |
| Math | Fractional Proportionality | Use in Note Values |
| Disc Golf | Use Putter Markers to Show Notes in Treble and Bass Clef | |
| Science | Wrapping Paper Roll Experiment | Frequency = Velocity / Wavelength |
| Enrichment | *Note Identification Activity* | |
| | Placing Colored Putters on Yarn Staff Lines | Spaces /Lines<br>Treble/Bass Clef<br>FACE EGBDF ACEG GBDFA |
| Technology | In Excel Create a Pie graph to show the Circle of 5th Chords | |
| Page 68 | *Anticipatory Vocabulary* | |
| Math | Proportion, Rate,<br>Rational Number | |
| Science | Pitch, Frequency, Amplitude, Wavelength, Waves, Crest, Trough | |
| Disc Golf | Speed, Glide, Turn, Fade | |
| Page 69 | *Math in Action* | |
| Make a Tree Graph | Color to Distinguish Note Values | Whole Note, Half Note, Quarter Note, Eighth Note, Sixteenth Note |

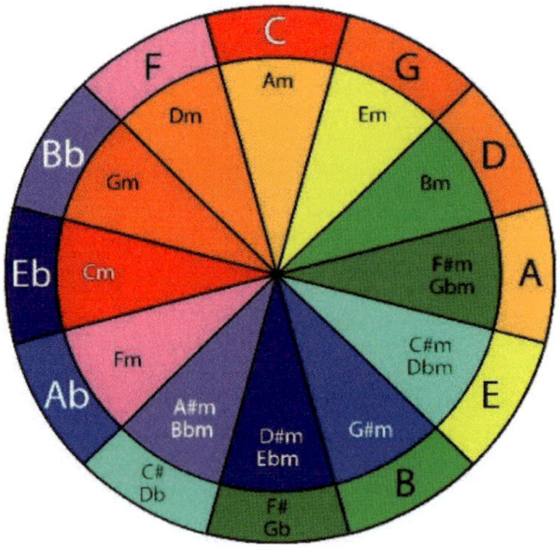

# LESSON #1

*(Idaho Common Core Standards in Italics)*

## Math Concept

Describing data sets using mean, median, mode, and range.

## Disc Golf Connection *(6-8.PE.3.1.2,3.1.3,3.1.4)*

Throwing Styles- Hyzer, Anhyzer, and Straight I(introduced on page 65)

**R**-right, **L**-left, **H**-hand, **F**-fore, **B**-back (RHBH/LHBH) (RHFH/LHFH)

## Math Connection *(6.SP.5.c,d)*

Students will record sets of straight throws by distance to calculate mean. They will also plot data on a bar graph to show relationships between median, mode, and range.

## Science Connection *(PS.1.3.1,3.2)(PS.2.A,B)*

Students will identify Newton's Laws of Motion. They will record data and show the variations in each law by calculating:

$F = m \times a$ $M = f/a$ $A = f/m$ $A = V_f - V_i/t$

## Enrichment *(6.NS.5.6.a)(7.NS.1.a)*

Using integers to describe scoring in Disc Golf.

A number's distance from 0 on a number line is its absolute value.

Place a piece of yarn between 2 chairs and fold 3x5 index cards in half. Have the students place the cards on the number line in numerical order using decimals, whole numbers, and fractions.

**Technology** Order data using an Excel sort command. *(6.NS.7.d.8)(6.SP.4)*

# 2

## Anticipatory Math Concepts

### *Data Set Terms*

***mean***- the sum of a set of numbers divided by the amount of numbers in the group

***median***- the middle or center number when the group is written in numerical order. When the data set has an even number of numbers, the mean is the average of the 2 middle numbers

***mode-*** the number that occurs most often. Some sets have no mode, or can have more than 1

***range***- subtract or find the difference between the smallest and largest number

## Anticipatory Science Concepts

### *Physics Terms*

***motion-*** the change in position of a moving object

***center of mass***- the point which acts as though the total mass of the object were at that position

***speed-*** the distance an object moves in a specific time interval

***velocity***- the direction or speed at which an object is traveling

***acceleration***- the change of speed of an object in a given time

***mass-*** the quantity of matter within an object

***inertia***- whether or not an object will resist a change in motion

***momentum***- the mass of an object multiplied by its velocity

### *Newton's Laws-*

1. If an object rests, or it has a consistency in speed or direction, the force on it is zero.
2. When the momentum of an object accelerates, then there is a separate force acting on it. Objects usually have a constant force that is proportional to their acceleration. The force and acceleration move in the same direction.
3. Forces occur in equal and opposite pairs called the action and reaction. When said object A exerts a force on object B, object B exerts an equal and opposite force on A. These forces are in balance.

## Anticipatory Disc Golf Concepts

### *Stability*

***stability-*** measuring a disc's ability to move side to side during its flight

***over-stable***- disc's flight moves to the left (for a right-handed, backhand throw - RHBT)

***under-stable-*** disc's flight moves to the right (for a right-handed, backhand throw - RHBT)

***turn-*** the disc's turning pattern from the start to the middle of its flight

***fade-*** the slowing down of the disc nearing the end of its flight

***rotation-*** the spin doesn't affect the lift and drag forces much, but determines a disc's stability while in flight

***spin rate-*** a disc's spin rate does not change while it loses velocity depending on its sloped design angle.

# DISC GOLF GAMES FOR FUN AND WARMUP

## Throwing Race

Pair the students into teams. Have students in horizontal rows facing each other, 10 yards apart. Let them throw their discs back and forth for one minute. Teams with the most catches in the air without dropping the disc, wins.

## Disc Flip

Students will pair off and toss the disc to land either on 'top' (label name up) or 'bottom' (opposite side). Toss the disc up like a coin. When it lands, the players side that is facing up must run to their 'home' spot. The opposing player has to catch them before they reach that space. You could use multiple teams and play Freeze Tag until everyone is caught.

## Activity Lab (Partners of 2)

1. Stand on the goal line of a football field or mark a chalk line on the ground.
2. Practice throwing using the correct techniques demonstrated by teacher/coach.
3. Using a stop watch, have a partner time how long a thrown disc was in the air from release to landing.
4. Measure the distance of the throw with a tape.
5. Throw five times to establish a mean or average and record the data in a notebook.
6. Once the data is collected, students will draw a bar graph to establish a visual display for mean, median, and mode.

## Math in Action

**How much force is needed to accelerate a 140 g disc at 5 $m/s^2$?**

What is known?

mass of disc, $m_d$ = 140 g

acceleration, a = 5 $m/s^2$

What is unknown?

force required, F

Choose the equation, F = ma

Solution:

The total mass, m, is 140 g

F = 140 g x 5 $m/s^2$

= 700 g · $m/s^2$ = 700 N **A force of 700 N is needed.**

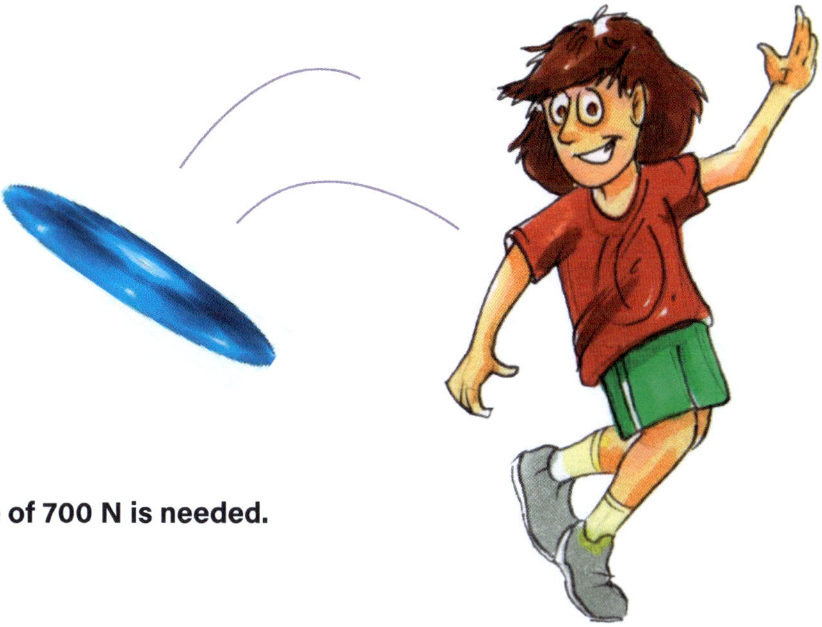

# 6

**A 75 kg disc golfer pushes off from a fence post with a force of 300 N. What is the disc golfer's acceleration?**

What is known?

mass of disc golfer, $m_{dg}$ = 75 kg

force, f = 300 N

What is unknown?

acceleration required, A

Choose the equation, A = f/m

Solution:

The total force, f, is 300 N

$A = 300\ N / 75\ kg = 4\ m/s^2$

**The disc golfer's acceleration is 4 $m/s^2$.**

# LESSON #2

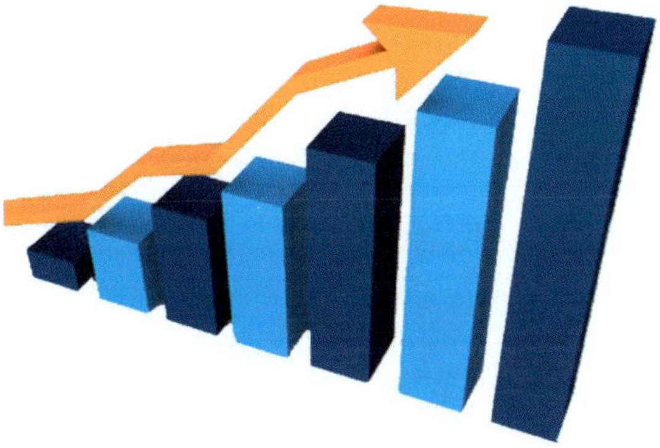

## Math Concept

Making and interpreting bar graphs and line graphs.

## Disc Golf Connection *(6-8.PE.4.1.1,4.1.2,4.1.3,4.1.4,4.1.5)*

Types of Discs: Putters, Mid-Range, and Driver

(RHBH/LHBH) (RHFH/LHFH)

## Math Connection *(6.SP.1.2.3)(6.SP.4)(6.SP.5.a,b,c,d)*

Students will record distances of throws by disc types in partnerships. They will plot data on bar graphs to show the variations of distances by disc type. Students will also connect data values using line segments in a line graph. In both types of graphs, students will draw, label, and choose appropriate scales to display the results of their data.

## Science Connection *(PS.2.MS.1,2)(PS.2.MS.4,5)*

Students will identify Newton's Laws of Motion. They will record data and show the variations in each law by calculating:

$D = v \cdot t$ $V = d/t$ $T = d/v$

## Enrichment *(6.NS.5-8)(7.NS.1-3)*

Using integers to describe scoring in Disc Golf. Disc Golf is scored much like regular ball golf. Par is simply a predetermined number of throws a disc golfer should require to complete a hole. If a disc golfer throws one over par, he scores +1, a Bogey. Two over par, +2, is a Double Bogey, and a Triple Bogey is +3. One under par, -1, is called a Birdie. Two under par, -2, is an Eagle or sometimes called a Double Birdie. Three under par, -3, is called an Albatros. Rarely does a golfer get -4, a Condor.

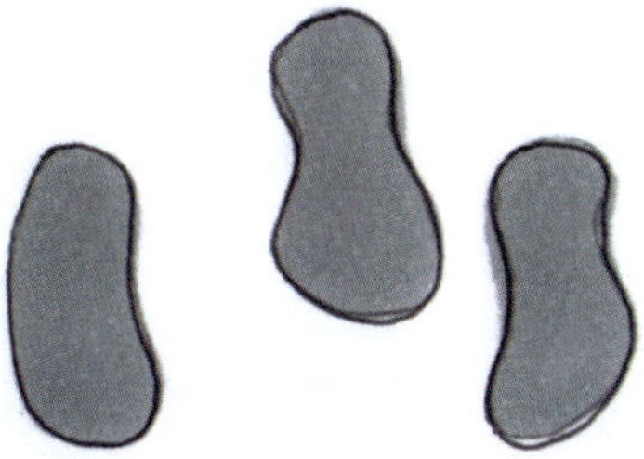

Let students cut out their own footprint, then use them as place holders to add and subtract integers as they move the prints up and down the number line on the floor.

For integers with similar signs, add their absolute values and use the similar sign.

For integers with dissimilar signs, subtract the smaller absolute value from the larger absolute value. When you have found the difference, place the sign of the number with the larger absolute value out front.

| | | | | | |
|---|---|---|---|---|---|
| Condor | -4 | Bogey | +1 | Par | 0 |
| Albatros | -3 | Double Bogey | +2 | | |
| Eagle (double birdie) | -2 | Triple Bogey | +3 | | |

**Technology** Using an Excel Spreadsheet to create Line and Bar graphs *(8.NS.1,2)*

## Anticipatory Math Concepts

### *Graphing Data*

***scales***- increments to show lengths of bars and lines numerically

***legends***- a display box that describe types of data used on the graph

***gridlines***- straight vertical and horizontal lines used to show variations in numerical values of the data displayed

## Anticipatory Science Concepts

### *Physics Terms*

***motion-*** the change in position of a moving object

***linear motion***: movement in a straight line

***speed-*** the distance an object moves in a specific time interval

***velocity***- the direction or speed at which an object is traveling

***acceleration***- the change of velocity of an object in a certain time

***Newton's Laws***-

1. If an object rests, or it has a consistency in speed or direction, the force on it is zero.
2. When the momentum of an object accelerates, then there is a separate force acting on it. Objects usually have a constant force that is proportional to their acceleration. The force and acceleration move in the same direction.
3. Forces occur in equal and opposite pairs called the action and reaction. When said object A exerts a force on object B, object B exerts an equal and opposite force on A. These forces are in balance.

## Anticipatory Disc Golf Concepts

### *Stability*

***putters-*** Discs created to fly straight, accurately, and predictably at slow rates of speed. They conform well with controlled shots that are near the basket and are used for short drives when shrubs, wooded areas, or other obstacles are in the players range.

***mid-range discs***- Discs that have sharper edges that allow them to cut through the air more easily. They generally travel faster, have more stability, and will fly further than a putter.

***drivers***- These discs have sharp angled edges and carry most of their mass towards the outer rim of the disc. The driver was created to travel longer distances at greater speeds of acceleration.

## DISC GOLF GAMES FOR FUN AND WARMUP

### Water Transport Game

Put water into the discs and make a mad dash for the finish line. Introduce relay teams with several members in each group.

### Disc Tag

Divide the students into multiple teams. The team that's ***not it*** will throw the disc back and forth trying to keep it away from the ***other*** team. Once a team captures the disc, they will switch places. You may wish to play it like Ultimate Frisbee with similar rules.

# Disc Golf Driver Practice

| THROWS | 1 | 2 | 3 | 4 | AVERAGE DISTANCE |
|---|---|---|---|---|---|
| Record Your Paces | | | | | |
| Instep | X | Paces | ÷ | 36 | |
| Record Your Yards | | | | | |
| TOTAL | Yards | ÷ | 4 | = | AVERAGE DISTANCE |

**Instructions:** Measure your instep from heel to toe in inches.

1. With a partner, stand approximately 40 yards apart.
2. Face each other and throw the disc.
3. Your partner will observe your throw and will stand behind where the disc lands.
4. Walk to your disc while counting the number of paces it takes to arrive where it landed.
5. Record your paces on your chart.
6. $\frac{\text{Insteps x paces}}{36} = \textit{total yards}$
7. Record your yards on your chart.
8. Calculate your average distance by adding your total distance in yards for all four throws and dividing by 4.

*Estimating Yards

1. Measure your instep in inches- (a normal stride or pace from heel to toe)
2. Multiply your instep times the number of paces you walked.
3. Divide by 36 to find the approximate (≈) numbers of yards you traveled.

***PLEASE BE MINDFUL OF OTHERS AROUND YOU AND THROW SAFELY!***

# Instructions for Creating Excel Spreadsheet

| | | |
|---|---|---|
| Cell A1 | **Heading: Disc Golf Driver Practice** | |
| Cell B2 | | Distance |
| Cell A3 | Toss 1 | type in distance of throws in yards* |
| Cell A4 | Toss 2 | |
| Cell A5 | Toss 3 | |
| Cell A6 | Toss 4 | |
| Cell A7 | Average | Add the sum of your four throws (yards) and divide by 4 |
| Insert: | Graphs- **vertical** (Drivers) | |
| Chart Icon | Appears in upper right hand corner of graph | |
| Green Plus + | Allows you to access: [right click to position any chart portals] (>) | |
| | Axes, Axes titles, Chart title, Data labels, Data tables, Gridlines, Legends | |
| Design | Allows for variation in background of chart- i.e. color, position of bars, 2-D, 3-D | |
| Color | Click on item you want to color and change colors with the Home icon A | |
| *Axes Titles* | *left side- [Distance in Flight] bottom- [Avenger SS 170 gr]* | |

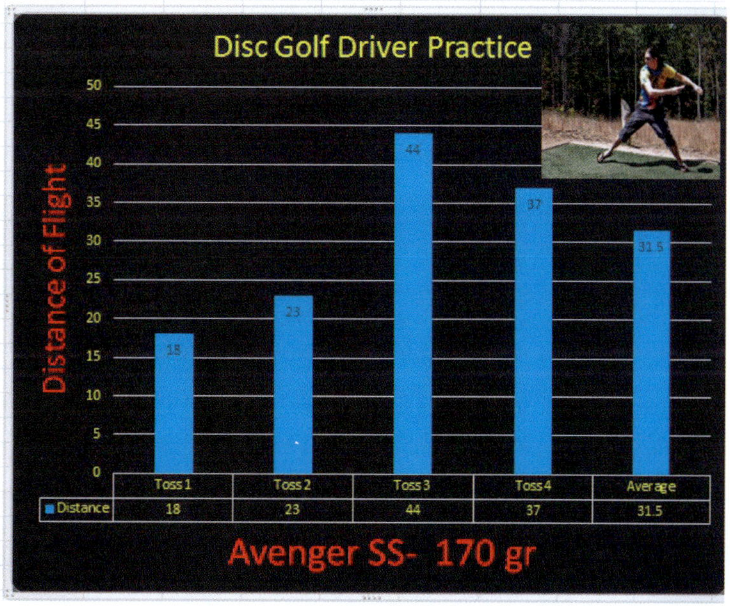


***Use contrasting colors for better readability- yellow lettering on black background.***

# Disc Golf Driver Practice

## Vertical Bar Graph Grading Rubric

| CATEGORY | 4 | 3 | 2 | 1 |
|---|---|---|---|---|
| **Data Labels** | Graph has an accurate title, a colored axis title on the left, and a colored axis title on the bottom of the graph. | Graph has a title and at least one colored axis title on left or bottom. | Graph has a title but no colored axes titles. | Graph has no title and no axes title. |
| **Bar and Table Labels** | Graph has vertical bars, tables, and table labels for each bar. Bars have different contrasting colors. | Graph has vertical bars but lacks table labels. Bars have different contrasting colors. | Graph has vertical bars, is missing tables and labels, and has no contrasting bar colors. | Graph has horizontal bars, no table or table labels, and has no contrasting bar colors. |
| **Mathematical Reasoning** | Chart shows thorough comprehension of the mathematical thought used to solve for the average distance thrown for 4 throws with 100% accuracy. | Chart shows adequate comprehension of the mathematical thought used to solve for the average distance thrown for 4 throws with 80% accuracy. | Chart shows basic comprehension of the mathematical thought needed to solve for the average distance thrown for 2 to 3 throws with at least 60% accuracy. | Chart shows minimal comprehension of the mathematical thought needed to solve for the average distance thrown for 1 to 2 throws with 40% accuracy or less |
| **Pictorial Evidence** | Used axes, axes labels, gridlines, and legends. | Used axes, axes tables, and labels. | Sometimes used axes, axes tables, and labels with no gridlines or legends. | Rarely used chart portals for axes, axes tables, labels, or gridlines and legends. |
| **Accuracy of Statistical Data** | All of the data was entered into the correct cells with 100% accuracy. | Only 80% of the data was entered into the cells correctly. | Only 60% of the data was entered into the cells correctly. | Less than 50% of the data was entered into the cells correctly. |

# Math in Action

**Your friend thinks he can run at a speed of 4 m/s. You are sure you can run faster. Your pal Jay times you as you run full blast for 100 m. You did it in 20 s. Who runs faster?**

What is known?

distance, d = 100 m

time, t, = 20 s

What is unknown?

velocity required, V

Choose the equation, V = d / t

Solution:

The total distance, d is 100 m

D = 100 m / 20 s = 5 m/s

**Your friend's speed is 4 m/s; you run faster.**

**Sound moves at a rate of 330 m/s. During a disc golf tournament, a lightning bolt struck the ground 1 km away from you. How long will it take for the sound to reach you?**

What is known?

distance, d, = 1 km

speed, v = 330 m/s

What is unknown? time required, T

Choose the equation, T = d / v

Solution:

The total distance, d, is 1 km

T = 1 km/330 m/s = 1000 m/330 m = 3.03 s

**It will take sound 3.03 s to reach you.**

# LESSON #3

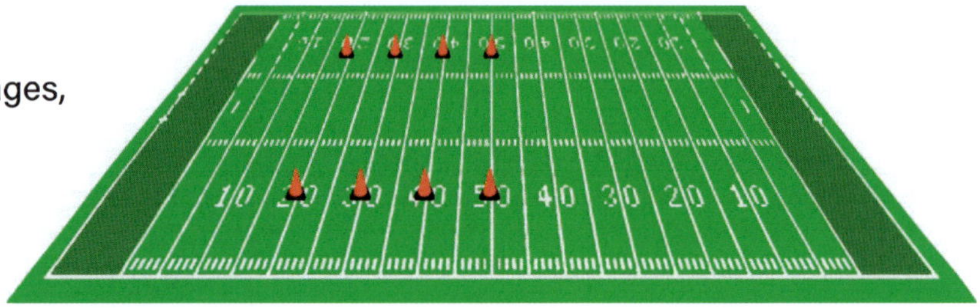

## Math Concept

Using Scatter Plots, Histograms, Percentages, Angle Sizes, and Classifications

## Disc Golf Connection

*(6-8.PE.2.1.1,2.1.2,2.1.3)*

Driving Practice: 1-step throw, 2-step cross and throw

(RHBH/LHBH)

## Math Connection *(7.G.2,5)*

Students will practice driving on a football field. They will record ten throws for distance and accuracy, then record the percentage of successful throws that land within 10 feet of the projected target.

Resting positions of all discs will be measured from points of origin and classified by their angle measurements according to the linear trajectory of their individual flight paths.

## Science Connection *(PS.MS.1,2,3,4,5)*

Student's will identify and describe important concepts related to gravitation:

- Newton's Law of Gravitation
- terminal velocity
- escape velocity
- free fall

## Enrichment

Using bar graphs to find constants of proportionality in percentages using means/extremes [k(x) = y] as well as angle identification. *(7.RP.2.a,b)*

Percentages

| 0 | 10 | 20 | 30 | 40 | 50 | 60 | 70 | 80 | 90 | 100 |
|---|---|---|---|---|---|---|---|---|---|---|
| 0 | 20 | 40 | 60 | 80 | 100 | 120 | 140 | 160 | 180 | 200 |

40% of 200 is 80 100x = 40(200) Value 80% of 200 = 160 100x = 80(200)

Angles that are smaller than 90 degrees are acute. Angles that are larger than 90 degrees are obtuse. Use a protractor to show no matter what way an angle faces, it's either acute or obtuse. Pay attention to the smaller and larger scales on the protractor. Always line up the bottom ray of the angle with the bottom of the line on the protractor, and make sure the vertex is in the center.

**Technology** Create a scatter plot or histogram to track distances in Excel Spreadsheet. *(8.SP.1)*

## Anticipatory Math Concepts

### *Graphing Intervals*

***frequency table-*** used in grouping intervals of numerical values

***frequency of an interval***- the amount of numbers appearing in an interval

***histogram***- a graph that shows numerical data in a frequency table. Individual bars are used to show each numerical interval. Frequency is used by showing varying bar heights.

***scatter plot***- numerical pairs on a graph shown as (x,y), which plots collective points in a coordinate plane

***acute***- an angle smaller than 90°

***obtuse***- an angle larger than 90°

## Anticipatory Science Concepts

***Newton's Law of Gravitation-*** states that there is a gravitational force of attraction between any two objects with mass, which depends on their masses and the distance between them.

***terminal velocity-*** the largest, constant speed reached by a falling body passing through a gas or liquid. As the speed increases, the resistance due to the air or liquid (drag) increases. Eventually, the weight will equalize or balance out.

***escape velocity-*** the smallest speed at which a body will travel in order to escape the gravitational tug of another body without any extended force.

***free fall-*** the unrestricted movement of an object when it is acted upon only by the single gravitational force where there are no resistive or other forces at play, like air resistance.

## Anticipatory Disc Golf Concepts *(6-8.PE.4.1.1)*

### *Rules of Disc Golf*

***teeing off-*** Play begins at the tee box. A player's foot must be in touching the tee box at the time they throw the disc.

***initiating position-*** After a disc is thrown, the exact area where it first comes to rest, when the disc has completely stopped all motion.

***maintaining the lie***- The original position where a disc lands after being thrown. A small mini disc marker is generally placed directly behind where the disc hit the ground after coming to a resting position.

***throwing from a fixed position***- While releasing a disc, a player must have one foot in direct contact with the playing surface on the lie.

***holing out***- When holing out on a basket target, the disc must come to rest within the bottom cylinder of the basket or inside of the chains.

***out of bounds***- When a disc crosses into an area that exceeds the perimeter of the designated perimeter of play. Penalty throws are charged when discs go out of play and can be thrown up to a meter from the previous lie and perpendicular to the fowl position.

***discs for play***- Playing discs must adhere to the conditions put forth by the PDGA Technical Standards. Any player that uses illegal discs will receive two penalty throws.

***using courtesy***- Players should and will not throw until they are cognizant that the thrown disc will not distract other players or injure anyone present. Littering on any disc golf course is discouraged and considered a courtesy violation. Players are encouraged to watch where other players discs land and search for lost discs when possible and plausible.

## DISC GOLF GAMES

### Hula Hoop Disc Throw

Attach several hula hoops onto a low-lying tree limb with string. Allow every player three or four chances to toss a disc through one of the hoops. The players may each gain a point each time a disc flies directly through the hoop. Hoops may be attached at lower or higher levels to increase the level of difficulty in the game.

### Castle Toss

Arrange a cone shape stack of cans on a table. Every participant will get several pre-determined turns to knock the cans from the table with a disc. The distance from the table can be altered for the level of difficulty within the game.

# Math in Action

**Sean was practicing his putting at school. He had 6 successful throws after putting 20 times. What were his percentages of successful and unsuccessful attempts?**

**Percent**

| 0 | 10 | 20 | 30 | 40 | 50 | 60 | 70 | 80 | 90 | 100 |
|---|---|---|---|---|---|---|---|---|---|---|
| 0 | 2 | 4 | 6 | 8 | 10 | 12 | 14 | 16 | 18 | 20 |

What % of 20 is 6? **Value** $20x = 6(100)$

$k(x) = y$ $k = 5$

30% were successful 70% were unsuccessful

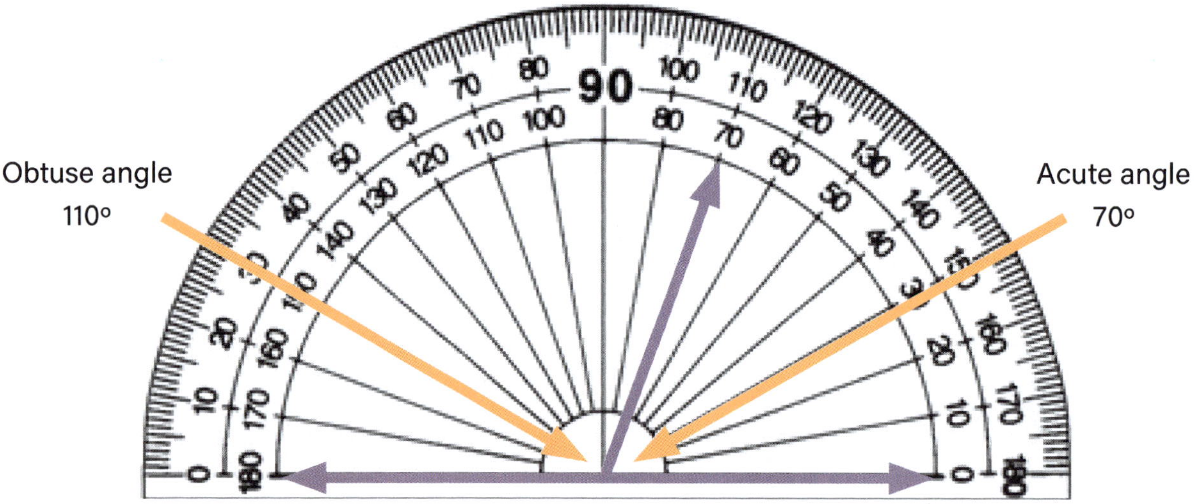

# LESSON #4

## Math Concept

Solving Equations

Stem and Leaf Plots

## Disc Golf Connection *(6-8.PE.2.1.1, 2.1.2, 2.1.3)*

Throwing- Using mid-range and driver disc types.

(RHBH/LHBH) (RHFH/LHFH)

## Math Connection *(6.RP.3.a)(6.SP.5.a,b)*

Students will use a circular tape measure to mark and label regular intervals with cones. Use a string to make a center line for aiming throws. They will practice aiming towards a target with mid-range or driver discs and record consistency by plotting their throws on a stem and leaf graph.

## Science Connection *(PS.2.MS.1,2,4,5)*

Students will identify the components of a simple machine, including:

levers (1st, 2nd, 3rd class) fulcrum resistance arm effort arm machine efficiency mechanical advantage work

## Enrichment *(PS.3.MS.1,2,3,4,5)(PS.3.A,B,C)(ETS.1.A,B)*

Using the following formulas, students will calculate the mechanical advantage using each component within the realm of a lever.

**$MA = L_e / L_r$ Effort arm ($L_e$) Resistance arm ($L_r$)**

## Technology *(6.SP.4)(7.SP.8.b)*

Create an Excel Spreadsheet to create a Stem and Leaf Plot Graph.

## Anticipatory Math Concepts

### *Graphing Data*

***scales***- increments to show lengths of bars and lines numerically

***legends***- can describe types of data used on the graph

***stem-and-leaf plot-*** distributing and arranging numerical data in a way that shows and demonstrates mean, median, mode, and range in a graphing format

## Anticipatory Science Concepts

### *Physics Terms*

***1st class lever-*** where effort and resistance forces are divided by a fulcrum that is used to multiply force

***2nd class lever***- when a resistance force is located between the effort force and the fulcrum that always multiplies force

***3rd class lever***- when an effort force is located between the fulcrum and the resistance force. The effort arm is shorter than the resistance arm, so it can't multiply force and the MA is always less than one

***lever-*** a free pivoting bar that can turn, attached to a definite fixed point

***fulcrum-*** the part of a lever that has a fixed point

***resistance arm***- the point on a lever where the resistance force is exerted

***effort arm***- the point on a lever where the effort force is applied

***machine***- a device that is used to supersede a force or (load)

***efficiency***- a percentage expressed by the work done on the load divided by the work done by the effort

## Anticipatory Disc Golf Concepts

***discs-*** Today discs are much smaller and heavier than traditional flying discs. They measure 8-9 inches (20-23 cm) in diameter and have a mass between 120-180 grams.

***features***- The design of modern discs are shaped for speed, precision, and control. They are categorized as: putters, mid-range discs, and drivers.

# DISC GOLF GAMES

## Disc Tic Tac Toe

Place a big tarp on the ground and make a tic-tac-toe board on it with dark masking tape. Let the team players take turns tossing marked colored discs onto the board. Players that land three discs in a row win the game. Students can practice defensive and offensive strategies to opposing player's tosses. You can also play in teams and create a Round Robin Tournament over a period of several days.

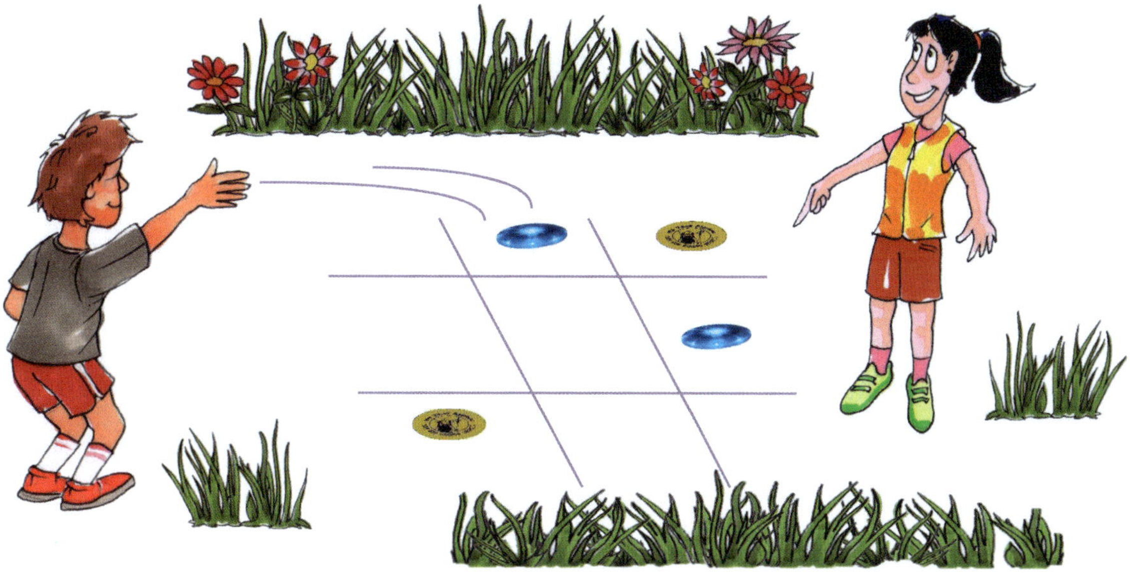

## Disc Bowling

Place soda cans or plastic bottles in lined groups for bowling pins and try to knock them down with the discs. Use the same scoring dynamics as regular 10-pin bowling. Distances from the pins can be determined by age level for difficulty of play.

# Math in Action

**Kim was constructing a tee pad for her school's disc golf course. She was digging and hit an old heavy backpack that was buried in the ground. She used a crow bar and applied an effort force of 30 N to extract it. The backpack had a resistance force of 1500 N. What was the mechanical advantage of the crow bar?**

What is known?

Resistance force, $F_r$ = 1500 N

Effort force, $F_e$ = 30 N

What is unknown?

Mechanical advantage MA

Choose an equation

$MA = F_r / F_e$

Solution:

Resistance force = 1500 N

1500 N / 30 N = 50 N

**She needed a mechanical advantage of 50 to extract the backpack..**

## Stem and Leaf Plots

James wanted to practice throwing his DX Archangel driver. He threw 17 times and needed to keep tract of his results. He listed his throws in a stem and leaf plot.

**Drives** **Key 32 | 1 = 321**

| Stem | Leaf |
|---|---|
| 32 | 1, 3, 5, 5, 5 |
| 33 | 2, 6, 7, 8, 8, 9 |
| 34 | 3, 4, 4, 5, 8, 9 |

Mode = 325 Range = 28

Median = 338 Mean = 336

# LESSON #5

## Math Concept

Circles: Circumference/Area

## Disc Golf Connection *(6-8.PE.1.1.1,1.1.2,1.1.3)*

Putting Practice

(RHBH/LHBH)

Staggered and Straddle Positions

## Math Connection *(7.G.4)*

Students will throw putts inside a circle from the center, towards an outside edge. They will throw from distances of 5, 10, 15, and 20 feet. They will putt from $45^\circ$, $90^\circ$, $135^\circ$, $225^\circ$, $270^\circ$, and $315^\circ$ angles, using the standard three stances for putting.

Describe the positions on a compass face to the degrees thrown from a direct point of origin. Students will determine and match the numerical coordinates for each of the 8 bearings on a compass. [North, East, South, West, Northwest, Northeast, Southwest, Southeast]

## Science Connection *(PS.2.MS.1,2,4)(PS.3.MS.1,2,3,4,5)*

Student's will identify and describe circular motion in terms of:

centripetal acceleration    centripetal force    centrifugal force    Newton's 2nd Law

$a = v^2/r$

## Enrichment *(7.G.2,5)*

Naming and identifying central angles, radius, diameter, and chords

## Technology *(7.RP.2.b)*

Creating pie graphs to display data on an Excel Spreadsheet

## Anticipatory Math Concepts

***circumference-*** the distance around a circle, as used in the formula $2(\pi)(d)$

***area***- the interior measure of a circle, as used in the formula $(\pi)(r)^2$

***radius***- any point on the perimeter of a circle and its distance to the center

***diameter***- connecting points on the perimeter of a circle that pass through the center

***central angle***- the angle whose vertex is located in the circle's center and whose endpoints touch the outer edge or circumference of the circle

***chord***- a straight line segment whose endpoints both lie on the circle, or join 2 points on any curve

## Anticipatory Science Concepts

***centripetal acceleration***- the acceleration of an object in circular motion acting towards the center of the circle. $A = v^2/r$ $v$ = velocity around the circle $r$ = radius of circle

***centripetal force-*** the force which acts on an object towards the center of a circle to produce centripetal acceleration, and so keeps the object moving in a circle

***centrifugal force-*** the equal and opposite reaction to the centripetal force. It does not act on the object moving in the circle and is not considered

***Newton's 2nd Law-*** When the momentum of an object accelerates, then there is a separate force acting on it. Objects usually have a constant force that is proportional to their acceleration. The force and acceleration move in the same direction.

## Anticipatory Disc Golf Concepts

***underhand putt-*** The player pushes the disc towards the basket from the solar plexus. Upon release of the disc, continue to point with your index finger towards the target. The more wrist you include, the greater the spin. The thumb should be pointing straight up like when shaking.

***overhand putt***- When a player holds a disc level, supported in the center by the thumb, with the finger tips beyond the edge, as if you were supporting a plate. The player extends his opposite leg away from the throwing arm, reaches back, and then points their arm towards the basket, tossing the disc in a fashion similar to when throwing a dart. Ideally the thrower wouldn't turn or rotate the wrist. A gentle follow through will generally give the disc its spin. Sometimes lifting the opposite leg will ensure a proper release.

# DISC GOLF GAMES

## Disc Horse

Instead of using a basketball, use discs for this game of HORSE. The initial player aims for a target or object in the area close by. Sometimes it's fun to use a regular basketball court. If the opposing player also hits your designated target, every player must try it. You must complete the leader's challenge, and if you fail to make the shot, you will receive a letter until you spell HORSE. The first person to miss 5 times is out. Proceed until all the players have missed 5 shots, then start a new game. Take turns being the challenger.

## Disc Bocce

Use a small mini marker and different colored drivers. Throw the mini marker approximately 5-10 meters out into the field or yard. Each player tosses their driver towards the mini marker. The player who tosses their colored driver closest to the mini marker gets 1 point. Toss the mini marker again and continue play until someone scores 20, to win the game.

## Disc Golf Putting Practice

| Students will putt 2 tosses from each angle and record the number of putts MADE in the basket. | Left 60° | | Center 90° | | Right 60° | |
|---|---|---|---|---|---|---|
| STUDENT NAME ____________ | 3 paces ≈ 10′<br>DISTANCES | | 5 paces ≈ 15′<br>DISTANCES | | Rotate to the right →<br>DISTANCES | |
| | 10′ | 15′ | 10′ | 15′ | 10′ | 15′ |
| Toss 1 | Practice<br>Record | Practice<br>Record | Practice<br>Record | Practice<br>Record | Practice<br>Record | Practice<br>Record |
| Toss 2 | Practice<br>Record | Practice<br>Record | Practice<br>Record | Practice<br>Record | Practice<br>Record | Practice<br>Record |

**Instructions: Begin with 2 practice throws from each angle and distance.**

1. Students will be put into pairs and given two disc golf putters.
2. Each pair of students will record the data for each putt made from each putting angle. (left- 60° center- 90° right- 60°)
3. No more than three pairs of students should be assigned to a disc golf basket.
4. One student is to throw at a time, as their partner records ONLY the putts (x) that are *MADE* in the basket.
5. Only after ALL students have thrown their putts from the ten foot putting locations can they retrieve the putters, rotate through, then return to their new putting location to throw from fifteen feet.
6. Once student number one has completed their putts, putter number two completes the activity.

# Make an Excel Graph

Cell A1 **Heading: Disc Golf Putting Practice**

Cell C2, D2 10 15

Cell A3 Left Toss 1 Type in successful throws numerically or 0 for miss

Cell A4 Left Toss 2

Cell A5 Center Toss 1

Cell A6 Center Toss 2

Cell A7 Right Toss 1

Cell A8 Right Toss 2

Cell 9 Blank

Cell A10 Left

Cell A11 Center

Cell A12 Right

Cell A13 Average/%

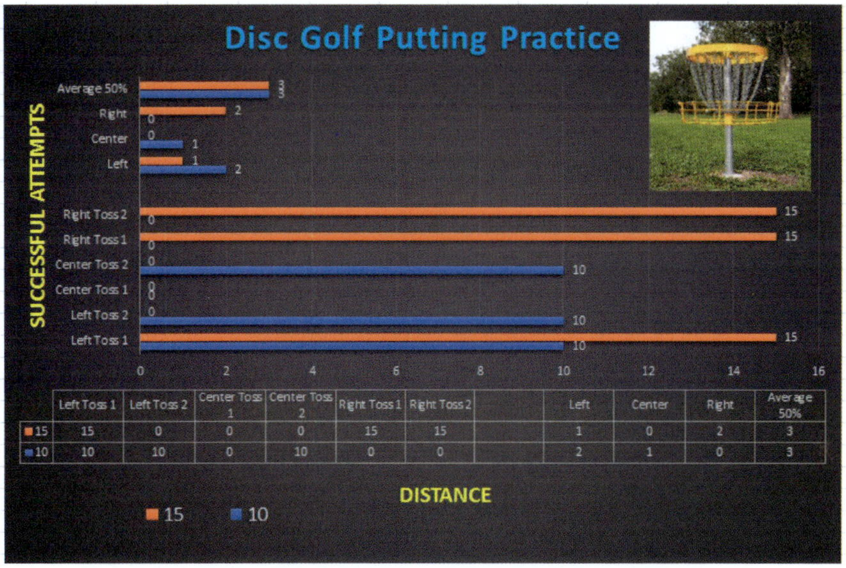


x/100 = (successful attempts) / (total throws)

Example: made 6 12x = 6(100)

Divide by 12 600 ÷ 12 = ***50%***

Insert: Graphs- **horizontal** (Putting)

Chart Icon Appears in upper right hand corner of graph

Green Plus + Allows you to access: [right click to position any chart portals] (>) shows options

Axes, Axes titles, Chart title, Data labels, Data tables, Gridlines, Legends

Design Allows for variation in background of chart- i.e. color, position of bars, 2-D, 3-D

Color Click on item you want to color and change colors with the Home icon A

*Axes Titles: left side- [Successful Attempts] bottom- [Distance]*

***Use contrasting colors for better readability- yellow lettering on black background***

# Disc Golf Putting Practice

## Horizontal Bar Graph Grading Rubric

| CATEGORY | 4 | 3 | 2 | 1 |
|---|---|---|---|---|
| **Data Labels** | Graph has an accurate title, a colored axes title on the left and one on the bottom of the graph. | Graph has title and at least one colored axes title on left or bottom. | Graph has a title but no colored axes titles. | Graph has no title and no axes titles. |
| **Bar and Table Labels** | Graph has horizontal bars, tables, and table labels for each bar. Bars have different contrasting colors. | Graph has horizontal bars and tables, but lacks table labels. Bars have different contrasting colors. | Graph has horizontal bars, is missing tables and labels, and has no contrasting bar colors. | Graph has vertical bars, no table or table labels, and has no contrasting bar colors. |
| **Mathematical Reasoning** | Chart shows thorough comprehension of the mathematical thought used to solve for percentage of successful shots. | Chart shows adequate comprehension of the mathematical thought used to solve for percentage of successful shot, but it is rounded incorrectly. | Chart shows basic comprehension of the mathematical thought needed to solve for percentage of successful shots, but solved percentage data incorrectly. | Chart shows minimal comprehension of the mathematical thought needed to solve percentage of successful shots by setting up percentage data incorrectly with incorrect solution. |
| **Pictorial Evidence** | Used axes, axes tables, labels, gridlines, and legends. | Used axes, axes tables, and labels. | Sometimes used axes, axes tables, and labels with no gridlines or legends. | Rarely used chart portals for axes, axes tables, labels. gridlines, and legends. |
| **Accuracy of Statistical Data** | All of the data was entered into the correct cells with 100% accuracy. | Only 80% of the data was entered into the cells correctly. | Only 60% of the data was entered into the cells correctly. | Less than 50% of the data was entered into the cells correctly. |

## Math in Action

**If the speed of a mid-range Stingray is 3 m/s, and it is traveling in a circular path equal to the radius of 2m, find the disc's exact centripetal acceleration.**

What is known?

velocity, v = 3 m/s

radius, r, = 2m

What is unknown?

acceleration, a

Choose an equation

$A_{centripetal} = V^2 / r$

Solution:

$V^2 / r$

$= 3\ (m/s)^2 / 2\ m$

$= 4.5\ m/s^2$ **The centripetal acceleration of the object is 4.5 m/s².**

# LESSON #6

## Math Concept

Rate    Unit Rate    Average Rate

## Disc Golf Connection *(6-8.PE.1.1.1,1.1.2,1.1.3)(6-8.PE.2.1.1,2.1.2,2.1.3)*

Throwing- Using putters across the width of a football or soccer field.

(RHBH/LHBH) (RHFH/LHFH)

## Math Connection *(6.RP.2.3.a,b,c,d)*

Students will practice throwing straight and angular throws across a field and record their results by accuracy in the disc's stability. Three cones will be set 10 m apart for the students to aim at. They will record and tract their percentage shots.

They can also use stop watches to calculate the average speeds of the discs in flight. Students will answer the question: Is there a relationship between speed, release, and accuracy?

## Science Connection *(PS.2.MS.1,2,4,5)(PS.3.MS.1,2,3,4,5)(PS.3.A,B)*

Students will list and identify the cause and effect ratios of:

lift    drag    gravity    air resistance    angular momentum

## Enrichment *(6.SP.4.5.a,b,c,d)*

Students will use rates and unit rates to evaluate and calculate average rate as distance divided by time.

## Technology *(7.RP.2.b)*

Create an Excel Spreadsheet to create a either a bar graph, line graph, histogram, frequency table, scatter plot, or stem-and-leaf plot to calculate correctly the percentage of successful or unsuccessful shots attempted.

## Anticipatory Math Concepts

***rate-*** the ratio of two related quantities that are measured in varying units

***unit rate-*** a ratio of two related amounts that combine to form a denominator of 1 unit

***average rate***- the rate or 'average velocity' of a moving object that will be found by dividing the distance traveled by the travel time

## Anticipatory Science Concepts

***lift-*** the force that allows a disc to stay in the air, as it opposes gravity's force against the mass of the disc in flight

***drag***- a backward force on the disc, as it goes against the disc's movement through the air

***gravity***- two objects with mass have a force of attraction between them that depends solely on their individual masses and the distance between each

***air resistance***- when an object's velocity increases, the resistance force increases as well due to the air or liquid (drag)

***angular momentum***- as a disc spins, it helps it fly by supplying angular velocity across its flight path to establish its stability by rotating. The faster the disc spins, the more stable it is

## Anticipatory Disc Golf Concepts

To compare and contrast various types of movement within a range of mathematical components that describe how a disc moves. Lift, drag, gravity, air resistance and angular momentum all play an integral part of how and why a disc moves the way it does. By showing proportional relationships of these factors, we can demonstrate how and why discs do what they do. (pg. 65 describes these affects in detail)

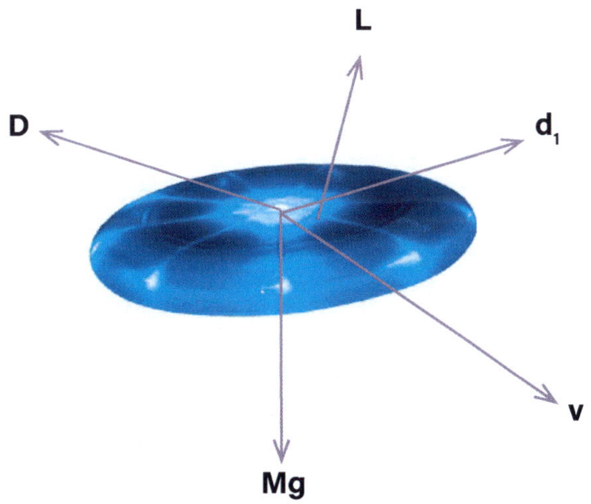


**v-velocity  mg-mass(gravity)  L-lift  $d_1$-direction of disc  D-drag**

# DISC GOLF GAME

## Disc Dodge Ball

Divide two opposing teams on opposite sides of a predetermined playing field. Chalk out a center line. Place the discs on the center line, two feet apart. Upon commencement of the game, each team's players run to the center line of the field to retrieve the discs. The object of the game is to hit opposing players below the waist safely. If a player is hit, they are called 'out.' Catching a disc in the air will allow one team member to come back into the game.

## Math in Action

**A Starfire distance driver was thrown, and it climbed 30 meters in 4 seconds. What was the disc's rate of ascent?**

What is known?

distance, d, = 30 m

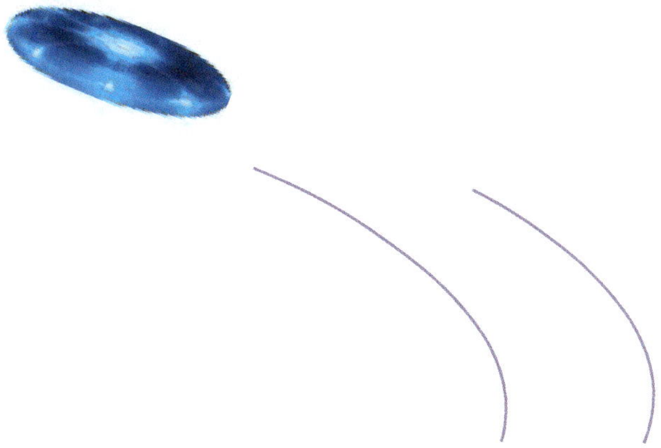

time, t, = 4 s

What is unknown?

rate, r

Choose an equation

R = d / t

Solution: distance = 30 m 30 m / 4 s = 7.5 m/s

**The disc's rate of ascent is 7.5 m/s.**

**Ryan and Jill were playing on a beautiful disc golf course in Hawaii. They walked 2.5 kilometers in one hour. How far was their walk that day if it took them 6 hours to complete the course?**

What is known?

rate, r, 2.5 km

time, t, 6 hrs

What is unknown?

distance, d

Choose an equation

D = r · t

rate is 2.5 km 2.5 km · 6 hrs = 15 km

**They walked 15 kilometers on the course.**

# LESSON #7

## Math Concept

Pythagorean Theorem

## Disc Golf Connection *(6.G.1,3)(7.G.1,2,3)*

Creating and constructing a Disc Golf Course

A landscaping team is putting in a disc golf park. One basket is placed in a corner where two walkways 15m x 17m intersect forming a 90° angle. Applying the Pythagorean Theorem, find the exact placement for the tee-off pad on the right.

## Math Connection *(8.G.6,7,8)*

Students will in-lay several wooden 4x4's at strategic places in the course between each basket, showing the exact location of each tee-off pad using the Pythagorean Theorem.

## Science Connection *(7.EE.3)(7.G.5,6)*

A rectangular Disc Golf course at your school has a length of 2400 ft. and a width of 1600 ft. John walks the diagonal distance across the course. About how far does he walk? Convert his distance to meters, if 1 meter = 39.37 in.

## Enrichment *(6.RP.2.3.a,b)(7.RP.2.a,b)(7.NS.2.c,d.3)*

Using a metric ladder, show conversions in the metric system for:

weight (grams)      capacity (liters)      distance (meters)

## Technology *(8.G.1.a,b,c.2-8)*

Create a graph on an Excel Spreadsheet to show missing mileage on parts of a road map, using the Pythagorean Theorem.

## Anticipatory Math Concepts

### *Triangles*

***legs***- the sides of a triangle that form angles

***hypotenuse***- the side of a triangle that is exactly opposite to the right angle

***Pythagorean Theorem-*** For all right triangles, adding the sum of the squares of the lengths of the legs is equal to the square of the length of the hypotenuse. ( $a^2 + b^2 = c^2$)

## Anticipatory Science Concepts

### *Metric System Terms*

***gram-*** used to measure mass or weight in the metric system

***liter***- used to measure capacity in the metric system

***meter***- used to measure length in the metric system

## Anticipatory Disc Golf Concepts

***targets-*** The Disc Pole Hole is often called the basket for general courses of disc golf.

***tee pads***- A starting pad is usually made on concrete and signifies where a player begins the hole. Mulch, decomposed granite, wood, or other natural materials are often found at the pads today.

***signage***- These are important in any and all disc golf course designs. They show accurate distances, boundary lines, par counts, and the layout of the entire course. They often identify mandatory paths, boundary lines, and lengths for the players to see clearly while they play.

***topography***- Disc golf is unique because it utilizes Mother Nature's wonderous design! Trees, shrubs, wooded areas, elevations changes, and creeks and ponds make each course unique, challenging and fun.

# DISC GOLF GAME

## Ultimate Disc Golf

Combine players into two teams of seven. Just as in the game of soccer, the strategy is to get the disc across the goal line successfully. Players must throw and pass the disc, transferring it up field to a teammate. Players must stop once they catch the disc. An unblocked disc over the goal line at no higher than shoulder height constitutes a point for the team.

# Math in Action

**Two student athletes were practicing at Hidden Valley Park. They wanted to become more accurate with their mid-range drivers. They had a math homework assignment to show how to convert distances using the metric system.**

Tina knew the names of the prefixes for the metric system, but was not sure how to move the decimal points when converting. Frank made some mistakes in his calculations. Can you find the errors below? Circle their mistakes, then make corrections if you dare!

**Remember: Grams = weight Liters = capacity Meters = distance**

| KILO | HECTO | DECA | | DECI | CENTI | MILLI |
|---|---|---|---|---|---|---|
| 50 | | | | 500,000 | | |
| .04 | | | 400 | | | |
| | 4505 | | | | .4505 | |
| | | 80.68 | | .8068 | | |
| | | | .00056 | | | 5.60 |
| | 3.2 | | | 32,000 | | |
| | | | .00093 | | .0093 | |

1. 5c cg = .0005 hg

2. 467 mg = 4.67 g

3. 150 dg = .0150 kg

4. 16.5 l = 16,500 cl

5. .0008 ml = 0.8 cl

6. 12 dkl = 1200 dl

7. 50 m = 50,000 km

8. .097 hm = 970 cm

9. 1560 cm = 1.56 dkm

## Disc Golf Complex Project

### Landscaping the Courtyard

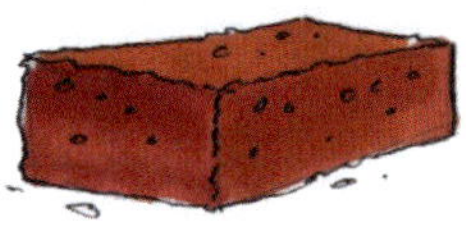

## Using Formulas for Geometric Shapes

| Trapezoids | Rectangles | Squares | Triangles |
|---|---|---|---|
| $\frac{1}{2}h(b_1 + b_2)$ | l(w) or b(h) | $s^2$ | $\frac{1}{2}$ b(h) |

# Designing a Disc Golf Complex Courtyard

You will create a Disc Golf Complex Courtyard using the polygons we have been using in class. Graph paper will be used to illustrate the complex and surrounding areas for the course. You will need to have grass, bark mulch, and cement pavers used in your project. As you design it, you need to think about which type of material you will need for each shape. After it its designed, you will need to find the total area of each shape as well as the final cost of the materials. *(6.G.1,3,4) (7.G.1,2,3,5,6)(8.G.1.2.a,b,c)(8.G.3,4,5)*

**1.** You must include 3 rectangles, 2 squares, 2 triangles, and 1 trapezoid.

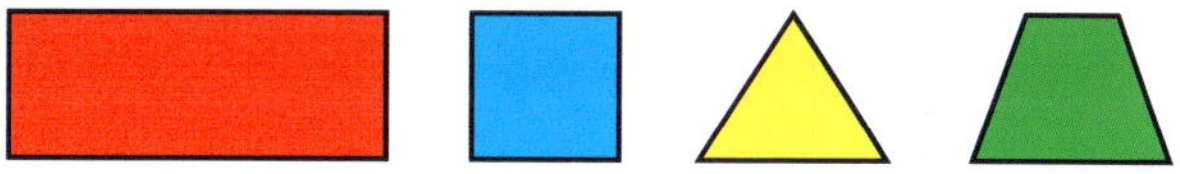

**2.** The 8 shapes must include at least one area of grass, cement pavers, and bark mulch.

- Cement pavers would cover patios, tee pads, walkways, barbeque pits, or under hot tubs
- Bark mulch would be to cover flower gardens, shrubs, or around trees
- Grass sod will cover the remaining area

**3.** Graph paper needs to be used for your final project. The length and width of each square grid box will equal one square foot. Please include dimensions for your grid for each geometric shape. *(7.RP.3)(7.G.1,2,3)*

**4.** The disc golf work page needs to be completed neatly and accurately.

**5.** Prices need to match the following: *(7.EE.3.4.a)*

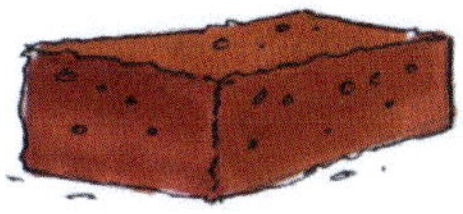

- Grass sod costs $0.25 per square foot
- Cement pavers cost $4.50 per square foot
- Bark mulch costs $0.30 per square foot

**6.** Final projects will be colored neatly to represent the actual complex courtyard area.

**7.** Two peers will edit your work upon completion.

## Math in Action

| Area of a Triangle | Illustration Examples |
|---|---|
| A triangle's area is half the product of the base and its corresponding height. | 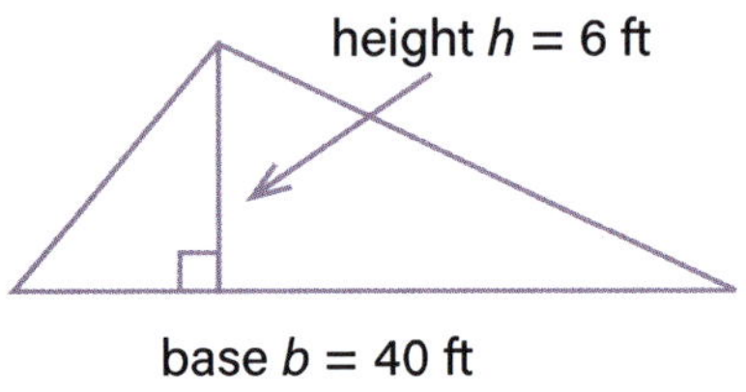  |
| A = ½ bh | A = ½ bh Write the triangle formula..<br><br>= ½(40)(6) Substitute 40 for b and 6 for h.<br><br>= 120 sq. ft. Simplify and solve. |
| **Area of a Trapezoid** | **Illustration Examples** |
| A trapezoid's area is half the product of the two bases and its height. | 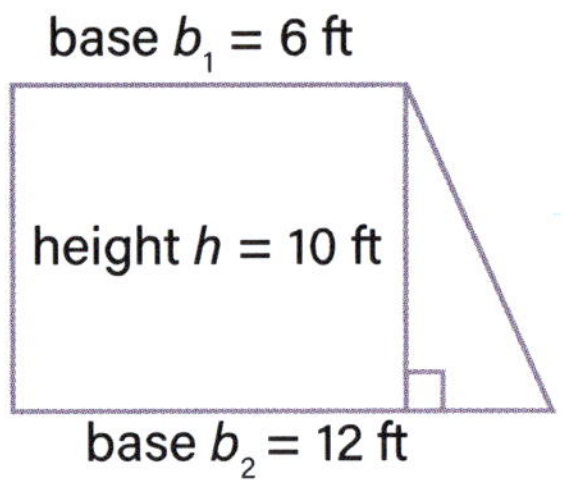  |
| A = ½($b_1$ + $b_2$)h | A = ½($b_1$ + $b_2$)h Write the trapezoid formula.<br><br>= ½(6 + 12)10 Substitute 6 for $b_1$, 12 for $b_2$, and 10 for $h$.<br><br>= 90 sq.in. Simplify and solve. |

# Disc Golf Complex Courtyard

| Shape | Location in Complex/Courtyard | Work (including formula) | Area | Work (price) | Price |
|---|---|---|---|---|---|
| **Example** | **Garden in Courtyard** | **A = l(w)<br>12(10)** | **A = 120 $ft^2$** | **Mulch<br>120(.30)** | **$36.00** |
| | | | | | |
| Rectangle<br>1 | l = ______<br>w = ______ | A = l(w) | A = _____<br>P = _____ | | |
| Rectangle<br>2 | l = ______<br>w = ______ | A = l(w) | A = _____<br>P = _____ | | |
| Rectangle<br>3 | l = ______<br>w= ______ | A = l(w) | A = _____<br>P = _____ | | |
| Square<br>1 | l = ______<br>w= ______ | A = l(w) OR $s^2$ | A = _____<br>P = _____ | | |
| Square<br>2 | l = ______<br>w= ______ | A = l(w) OR $s^2$ | A = _____<br>P = _____ | | |
| Triangle<br>1 | b = ______<br>h = ______ | A = ½ bh | A = _____<br>P = _____ | | |
| Triangle<br>2 | b = ______<br>h = ______ | A = ½ bh | A = _____<br>P = _____ | | |
| Trapezoid | b = ______<br>h = ______ | $A = ½(b_1 + b_2)h$ | A = _____<br>P = _____ | | |

| Grass<br>(Total courtyard minus other shapes) | Bark Mulch<br>(Find the sum of the shapes w/mulch) | Cement Pavers<br>(Find the sum of the shapes w/pavers) |
|---|---|---|
| Total cost of the grass =<br>$__________ | Total cost of mulch =<br>$___________ | Total cost of pavers =<br>$ __________ |

# Disc Golf Complex Courtyard

## (Extra Shapes- choose as needed)

| Shape | Location in Complex/Courtyard | **Work** (including formula) | Area | **Work** (price) | Price |
|---|---|---|---|---|---|
| **Example** | **Garden in Courtyard** | **$A = l(w)$<br>12(10)** | **$A = 120\ ft^2$** | **Mulch<br>120(.30)** | **$36.00** |
| Rectangle<br>1 | l = _______<br>w = _______ | $A = l(w)$ | A = _____<br>P = _____ | | |
| Rectangle<br>2 | l = _______<br>w = _______ | $A = l(w)$ | A = _____<br>P = _____ | | |
| Rectangle<br>3 | l = _______<br>w= _______ | $A = l(w)$ | A = _____<br>P = _____ | | |
| Square<br>1 | l = _______<br>w= _______ | $A = l(w)$ OR $s^2$ | A = _____<br>P = _____ | | |
| Square<br>2 | l = _______<br>w= _______ | $A = l(w)$ OR $s^2$ | A = _____<br>P = _____ | | |
| Triangle<br>1 | b = _______<br>h = _______ | $A = \frac{1}{2} bh$ | A = _____<br>P = _____ | | |
| Triangle<br>2 | b = _______<br>h = _______ | $A = \frac{1}{2} bh$ | A = _____<br>P = _____ | | |
| Trapezoid | b = _______<br>h = _______ | $A = \frac{1}{2}(b_1 + b_2)h$ | A = _____<br>P = _____ | | |

# Disc Golf Complex Courtyard Grading Rubric

| CATEGORY | 4 | 3 | 2 | 1 |
|---|---|---|---|---|
| **Trapezoid area/perimeter** | Work shows thorough comprehension when using variables to solve the trapezoid formula(s). | Work shows adequate comprehension when using variables to solve the trapezoid formula(s). | Work show basic comprehension when using variables to solve the trapezoid formula(s). | Work shows minimal comprehension when using variables to solve the trapezoid formula(s). |
| **Triangle area/perimeter** | Work shows thorough comprehension when using variables to solve the triangle formula(s). | Work shows adequate comprehension when using variables to solve triangle formula(s). | Work shows basic comprehension when using variables to solve the triangle formulas(s). | Work shows minimal comprehension when using variables to solve the triangle formula(s). |
| **Rectangle/ Square area/perimeter** | Work shows thorough comprehension when using variables to solve the rectangle/ square formula(s). | Work shows adequate comprehension when using variables to solve the rectangle/ square formula(s). | Work shows basic comprehension when using variables to solve the rectangle/ square formulas(s). | Work shows minimal comprehension when using variables to solve the rectangle/ square formula(s). |
| **Pricing using cost per sq.ft.** | Uses complex and refined mathematical reasoning in determining cost per sq. ft. | Uses effective mathematical reasoning in determining cost per sq. ft. | Some evidence of mathematical reasoning in determining cost per sq. ft. | Little evidence of mathematical reasoning in determining cost per sq. ft. |
| **Graphing a floor plan for a complex courtyard** | The work is displayed in a concise, defined way that easy to see and understand. | The work is displayed in an adequate way that shows understanding. | The work is displayed but has some details that are not so easy to understand. | The work appears disorganized and is difficult to understand. |
| **Artistic ability and neatness** | All drawings and pictures are outlined in black and neatly arranged with a professional appearance. | Most drawings and pictures are outlined in black and arranged neatly with adequate order. | Some, but not all drawings and pictures are outlined in black and are arranged with a basic order and appearance. | Drawings and pictures are not outlined in black and lack clarity and uniformity in appearance. |
| **Geometric accuracy** | Correct terminology and notation are always used for dimensions, area, and perimeter measurements. | Correct terminology and notation are usually used for dimensions, area, and perimeter measurements. | Terminology and notation used are inadequate, and it's not easy to understand the dimensional data for area and perimeter. | There is little or no use of appropriate terminology and notation with no accurate measurement in place. |

# LESSON #8

## Math Concept

Construct and design a model and a diagram of the muscle groups.

## Disc Golf Connection *(6-8.PE.2.1.2,2.1.3)*

List, label, and identify the minor and major muscle groups of the human body used during disc golf warm up calisthenics.

## Math Connection *(8.EE.5)(8.G.3,4)*

Create a diagram to chart the minor and major muscle groups of the human body.

## Science Connection *(PS.3.MS.1,2,3,4,5)(PS.3.A,B,C)*

Students will identify muscle groups of the upper, mid, and lower torso used in exercising.

## Enrichment *(PS.3.MS.1,2,3)*

Using a certified Red Cross blood pressure cuff, students will take three blood pressure readings before (pre-resting), during (moving), and after (post-resting) doing three exercises from the upper, mid, and lower torso. They can also measure heart rate per minute with a stethoscope. They will calculate the differences in the body's reaction to the influence of exercising to determine the benefits and effects of a healthy life style. They will repeat the program for three weeks to determine the numbers for exercising all three sections of muscles and graph them on a line graph.

They may also look for the percentage rate-of increase for success, using a relative frequency table to chart progress. The same may be done for resting, moving, and post heart rates as well.

## Technology *(6.RP.3.d)(6.SP.4.5a,b)(8.F.4.5)*

Using an Excel Spreadsheet to create Line Graphs, Bar Graphs, and Relative Frequency Tables students will show and identify ideal blood pressures and heart rates (pre, during, and post) while exercising.

## Exercising Muscle Groups of the Human Body

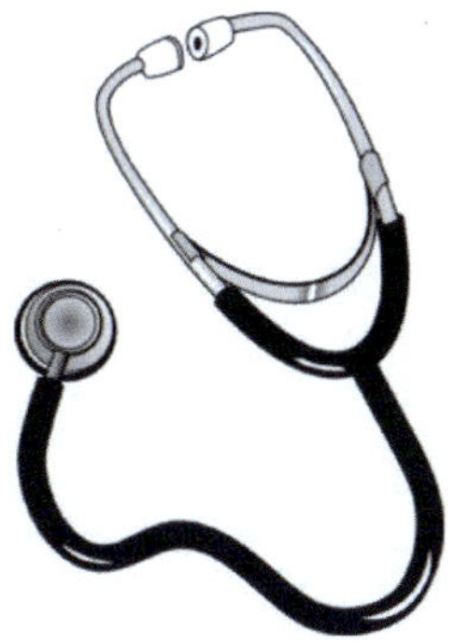

### *Upper, Mid, and Lower Torso*

**Upward Caterpillar**

abdominal muscles

**Downward Caterpillar**

gluteus (maximus, medius, and minimus), hamstrings, gastrocnemius, and deltoids

**Hurdler's Stretch**

hamstrings

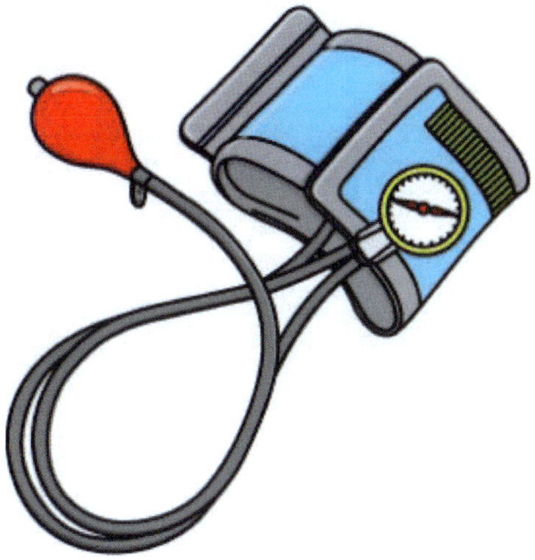

**Butterfly**

abductor muscles of the upper thigh

**Spiral Football Stretch**

gluteus muscles of the hip and buttocks

**Standing Quad Stretch**

quadriceps of the legs

**Lateral Shoulder Stretch**

deltoids and infraspinatus muscles of the shoulder

**Back Scratchers**

triceps muscle of the arm

**Lunges**

gastrocnemius and soleus muscles of the calf and Achilles tendons muscles of the ankle

## Frontal Muscle Anatomy

## Warm Up Calisthenics *(6-8.PE.1.1.1)(6-8.PE.2.1.2)*

### Upward Caterpillar

- Lie flat on the ground with your hands down by your sides.
- Draw your legs together and press the tops of your feet into the ground using your back, and gently lift your chest off the ground.
- As you stretch up inhale, as you lift your chest and stretch, exhale.

### Downward Caterpillar

- Begin on all fours with your hands directly under your shoulders and knees under your hips.
- Make sure your hands and feet are approximately the same width apart. Curl your toes under and extend your hips upward, making your body into a V shape by extending your shoulders away from your ears.
- Inhale as you curl your toes in, shifting your waste upwards and exhale as you press your fingers transferring your weight back down.

### Hurdler's Stretch

- Sit while extending a leg outward and keep the other leg bent at the knee.
- The heel should be placed to the inside of the knee.
- While inhaling, lower your thigh and calf towards the ground.
- Make sure your extended leg is straight and stretch the upper torso while exhaling.
- Switch legs and repeat the exercise.

### Butterfly

- With your heels touching, sit with the hips facing outwards.
- While grabbing your feet or ankles, pull them towards yourself, inhaling.
- Place the elbows near the inner thighs or knees and push your legs at the ground while exhaling.
- Remember to keep your back as straight as possible during the exercise with head up.

**Spiral Football Stretch** *(6-8.PE.3.1.1)*

- Sit with your legs extended and your hands behind your hips.
- Cross your left foot over your right ankle and point your heel towards yourself while inhaling.
- Put your right elbow to the right of the left knee while exhaling.
- Flex your hips while looking over your left shoulder, pushing down on your knee with your right elbow.
- Do this with both sides of your body.

**Standing Quad Stretch**

- Flex a knee outward and raise your heel towards yourself.
- Slightly flex, while exhaling and support your leg as you grasp your raised foot.
- Inhale and slowly pull your heel toward the rear.
- The inside of the legs should be touching while your pelvis rotates backwards.
- Switch legs and repeat the exercise.

**Lateral Shoulder Stretch**

- Stand with one of your arms raised to shoulder's height.
- Flex and pull your arm near the elbow across to the opposite shoulder, while inhaling.
- Grab your extended elbow with the other hand, pulling your elbow backwards while exhaling.

**Back Scratchers**

- Stretch your arm and extend it over your back and touch your hand to your shoulder blade while inhaling.
- Grab your elbow with the opposite hand, and pull it behind your head while exhaling.

**Lunges**

- Stand with your feet and hips width apart with your knees slightly bent.
- With your back straight and erect, relax and inhale.
- Take big steps forward, backward, sideways, and diagonal, and alternate your lead leg while exhaling.
- Karaoke can be done while sidestepping also, by extending the lunge to alternating the legs in front and then in back while sidestepping with your arms extended horizontally for balance.
- Repeat exercise for each direction, remembering to keep your knees from going past the outstretched foot.

## Exercises

**Reverse Crunches** *(6-8.PE.5.1.1)*

- Lie on the ground on your back, with your knees bent and hands behind your head.
- Lift the legs to a 90º angle.
- Keep a space between your head and your chin and chest, and look at the sky while inhaling
- Exhale and pull your knees towards your chest, then inhale and lower your legs back to your starting position.

**Hip Extensions**

- On your hands and knees, place hands shoulder width apart.
- Extend your right leg straight back and stretch outward.
- Keep your leg lifted and lower your chest to the ground until your elbow are at a 90º angle, then push upwards.
- Repeat exercise for several repetitions on each leg.

**Push Ups**

- Lie on the ground with palms down and toes pointed down.
- Lift body up to an upright horizontal position, arms extended with toes down.
- Inhale while you go down.
- Exhale while you come up.
- Repeat exercise for several repetitions.

**Squat Thrust**

- Begin in a vertical standing position.
- Crouch to a baseball catcher's squat with back and head straight with arms outstretched.
- Thrust forward, palms down on contact, while kicking legs outward towards the rear to a horizontal position.
- Recoil back to the catcher's position.
- Thrust entire body back to a standing position.
- Repeat exercise several times.

**Jumping Jacks**

- Start in a standing position, feet together, arms to the sides.
- Jump outwards spreading the legs about 2 feet apart while thrusting hands upward and clap above your head.
- Return to the original vertical position and repeat several times.

### Three-finger Grip

3 finger

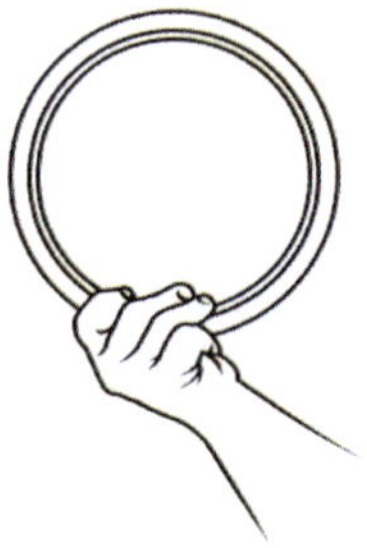

- Put your fingers on the inside rim of the disc.
- Fingers should be curled underneath the rim.
- Your index finger should not be placed on the exterior portion of the disc rim.

### Three-finger Grip Variation- (more stability)

- Place your index finger away from the ring an middle finger slightly.
- The index fingers is still places on the inside rim of the disc, but the edge of the disc lip will lie perfectly in the first crease line of the index finger.
- More grip stability, more throwing accuracy, sacrifice of distance.

Variation

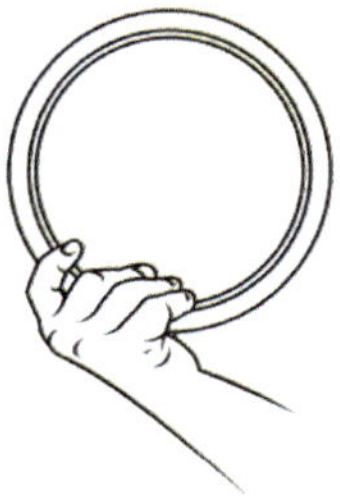

### Power Grip

- Pinky finger is placed on the inside lip of the disc.
- You can place your fingertips, so they are touching the corner of the inside lip where the flight plate of the disc meets the rim.
- You can place fingertips on the inner edge of the rim, so the tips are not actually touching the inside corner or the flight plate of the disc.

Power

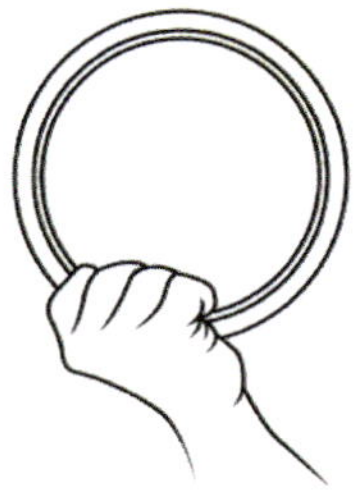

### Thumb on all Three Grips

- Lay the thumb completely flat.
- Flex the thumb, so only the tip is touching the flight plate of the disc.
- Leave the lower part of the thumb in the air.

*Maximum Distance*

- Keep thumb near the outer portion of the rim.

*Increased Stability*

- The farther you move the thumb toward the center, the more stable the throw.

Basic Putting Grip (forehand)

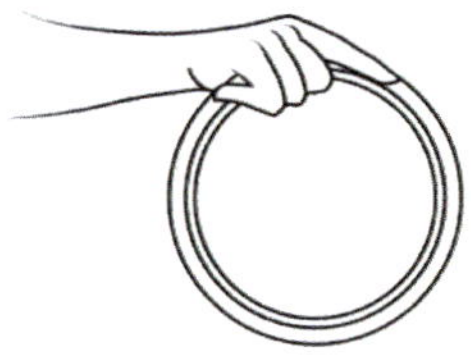

### Fan Grip

- Separate the fingers out to the underside of the flight plate of the disc.
- Place the index finger firmly on the rim so that the disc rests in the first crease line.
- Tuck the index finger around rim and fan middle ring finger on flat part of disc.
- Make sure the pinky finger does not touch the disc.

Fan

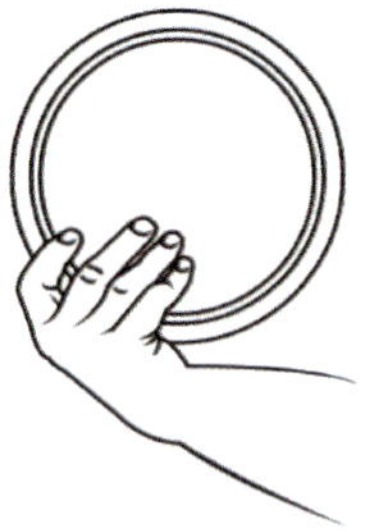

### Split Finger Grip

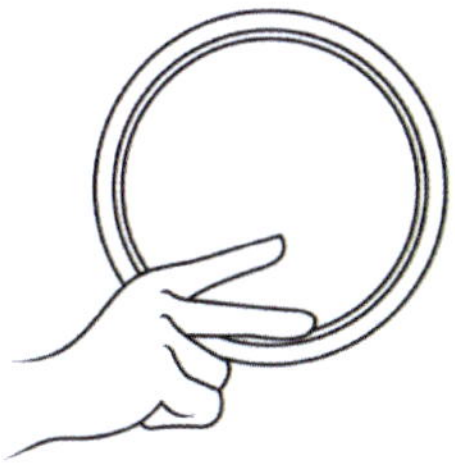


- Place the index finger on the inner flight plate of disc as it points towards the middle of disc.
- The middle finger is then split from the index ginger, creating V shape with the two fingers.
- Make sure the fingertip of the middle finger is placed on the inside of the disc rim.

### Stack Grip

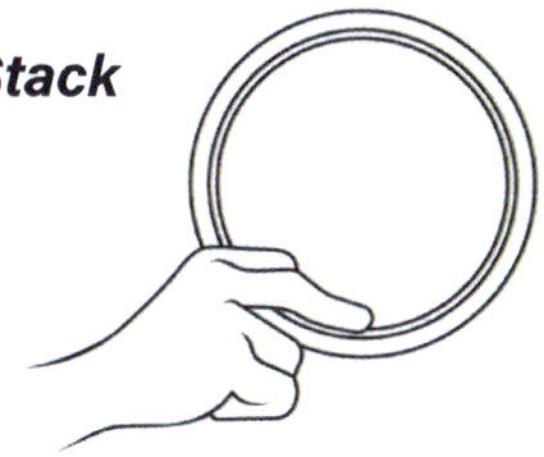


- Place your middle finger firmly on the inside and partially on the flight place.
- Place your index finger closely beside your middle finger to create a unified feel.
- Place both fingers side by side on the inner rim of the disc so they are pointing in the same direction.
- For more leverage, bend the middle finger inward at the first or second finger digit, so there is more tension put on the rim of the disc.

### *Thumb Position*

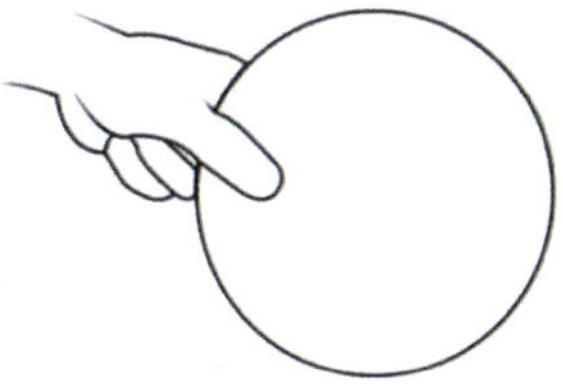

- It is always placed on top of the disc for stability.
- The middle fingertip should straddle the interior disc rim and the inner flight
- The ring and pinky fingers are used for support and stability.
- The exterior rim of the disc should be positioned in the corner of skin at the base of the index finger and thumb.

# LESSON #9

## Math Concept

Construct a tree diagram to show the sample space for a healthy meal.

## Disc Golf Connection *(7.RP.2.a,b,c)*

At each hole, identify and list a fresh fruit, starch, animal protein, nut/seed, and vegetable to make a meal. They must agree with the percentages for fats, carbs and proteins on the nutrition wheel.

## Math Concept *(7.SP.8.a,b,c)*

Students will create a sample space of foods that when mixed together correctly, will form a tree diagram that creates a daily meal plan that meets the FDA Standards for a healthy, well-balanced diet.

## Science Connection *(8-9.PS.1.1.1, 1.2.1, 1.2.2)*

List a complete daily diet plan, including the accepted amounts of foods in the food groups for a 2000 calorie diet. List by calorie the various foods from the nutrition pyramid. Percentages must meet the accepted levels for fats (20-35%), carbs (45-65%), and proteins (15-25%) allotted by the American FDA Standards.

## Enrichment *(6-8.H.5.1.6,5.1.7)(6.H.7.1.4)(6.RP.3.a,b,c,d)*

Students will create a weekly diet plan as prescribed by the FDA for a 2000 calorie a day diet. Categories for each of the food groups from the food pyramid must be met for daily food intake. All foods will be placed on a pie graph and must meet the criteria for healthy eating habits according to the carbs, fats, and protein percentages.

## Technology *(6-8.H.7.1.1, 1.2, 1.3)(8.SP.1,2,3,4)*

Students will create a Regression Analysis Chart in Excel to show the variations between certain food groups, to maximize the potential of combining specific fruits and vegetables to create a balanced diet.

## Anticipatory Math Concepts

***population***- an entire group of people or objects about which you want information

***sample space***- the part of a population you are drawing conclusions about or testing

***tree diagram***- a diagram in a tree shape that shows the possible outcomes or probable choices that happen in a sample space

## Anticipatory Science Concepts

***fats***- a major fuel source and storage form of energy in the human body

***carbohydrates***- any of the fibers, starches, or sugars that are found in vegetables, grains, fruits and milk products. They contain carbon, hydrogen and oxygen and provide energy for the human body.

***proteins***- one or more chains of amino acids in large molecules required for the regulation, structure, and function of the body's organs, tissues, and cells

## Anticipatory Disc Golf Concepts

### *Cause and effects of wind resistance*

***joule***- A unit of work and energy. 1 joule = a force of 1 newton moving through 1 meter.

***Law of conservation of energy***- Energy can neither be created or destroyed. It can change forms.

***Law of conservation of momentum***- A group of objects' total momentum is conserved unless a net force acts on them.

***mechanical advantage***- The number of times a machine multiplies the effort force applied to it.

***mechanical energy***- The total amount of kinetic energy and potential energy in a system.

***resistance force***- The force exerted by a machine to overcome resistance to gravity or friction.

# Combining Food Groups to Make Meals

**Choose one category at each meal.**

| Fresh Fruit | Starches | Animal Protein | Nuts, Seeds, Dried Fruit |
|---|---|---|---|
| apples, avocados<br>bananas, berries<br>grapefruit, grapes<br>kiwi, mango<br>melons,<br>nectarines, oranges<br>papaya, peaches<br>pears<br>pineapple | acorn squash<br>avocados<br>butternut squash<br>beans & lentils<br>cooked corn<br>millet, spelt<br>oats, buckwheat<br>pasta & cereals<br>quinoa & amaranth<br>sweet potatoes & yams<br>wheat, barley, & rye<br>white potatoes | red meat<br>pork<br>poultry<br>fish<br>milk<br>cheese<br>yogurt<br>ice cream<br>eggs | almonds, bananas<br>brazil nuts, cashews<br>chia & flax seeds<br>dried fruit<br>dried coconut<br>hazel nuts, hemp seeds<br>macadamia nuts<br>mature coconut<br>pecans<br>sesame seeds<br>sunflower seeds<br>walnuts |

**Fill the rest of your plate with cooked or raw non-starchy vegetables.**

| | | | |
|---|---|---|---|
| bock choy<br>broccoli<br>cabbage<br>cauliflower<br>kale<br>leafy lettuces<br>spinach | beets<br>carrots<br>celery<br>cucumbers<br>green beans<br>onions<br>spaghetti squash | artichokes<br>fennel<br>raw fresh corn<br>scallions<br>sprouts<br>yellow summer squash<br>zucchini | asparagus<br>bell peppers<br>garlic<br>ginger<br>green beans<br>jalapenos<br>sugar snap peas |

**You can also enjoy the following neutral items: butter, cold-pressed oils, lemons, lime, coconut water, non-dairy nut or seed milks, heavy cream, 70% or darker chocolate**

**Wait 3 to 4 hours before switching categories**
***(If you become hungry between meals, snack on non-starchy vegetables as desired.)***

| | | | | |
|---|---|---|---|---|
| Fresh fruit is best eaten alone on an empty stomach. It can also be used in salads or smoothies without impending digestion. | Bananas can be combined with fresh fruits OR with nuts, seeds, and dried fruits. | Beans and lentils are naturally more difficult to digest but can be considered a starch for our purposes. | Avocado can be combined with starches OR with fresh fruit OR with dried fruit. | Peanuts and soy nuts are not recommended due to their potential exposure to pesticides and mold. |

## Math in Action

What is known?

We don't always eat healthy percentages of the right foods that our bodies need to function properly.

What is unknown?

What percentage of foods should we eat and which foods should we limit or avoid?

Choose a healthy plan.

Stay away from consistently eating foods from the yellow, orange, and red groups.

Solution

Consistently choose several foods from the green and blue groups each day. Form dietary habits that include these 2 groups to live a healthier style of eating for life.

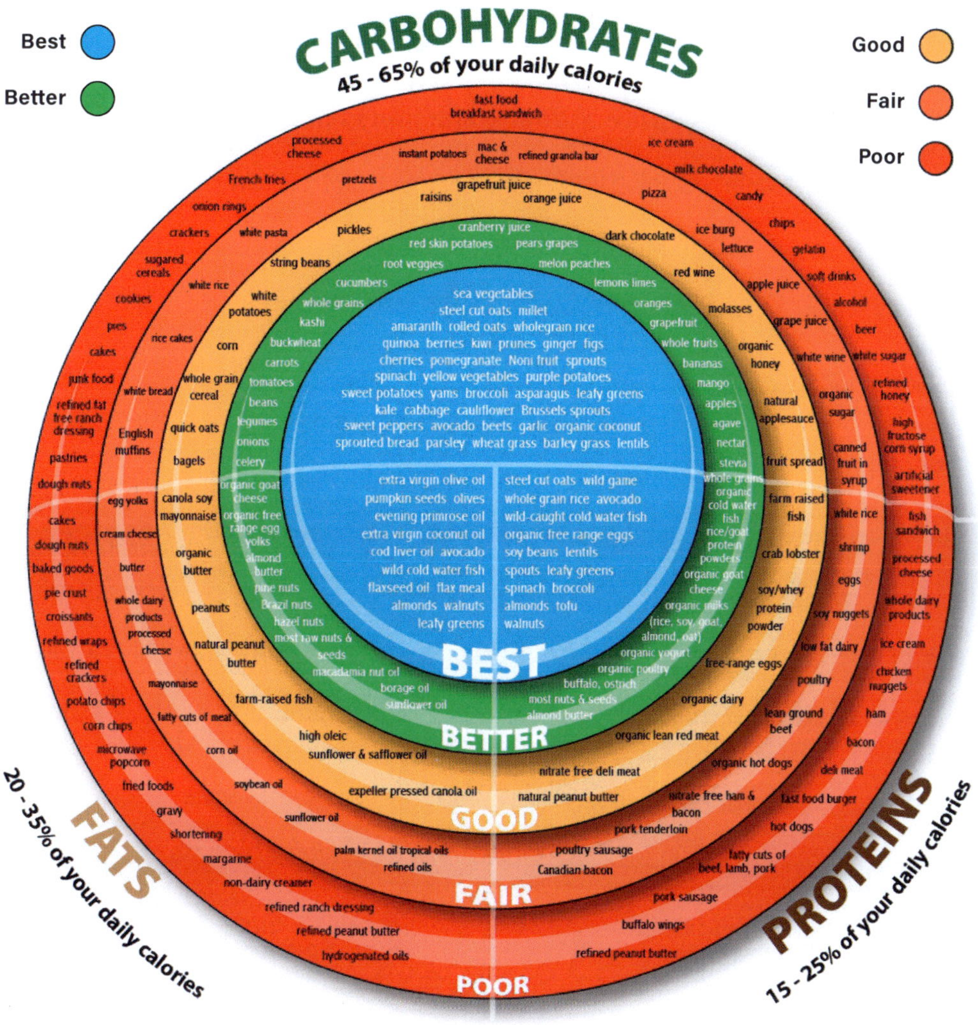

# LESSON #10

## Math Concept

Dimensional Analysis

## Math Connection *(6.RP.3.d)(6.NS.1)*

Adding and multiplying mixed numbers and fractions in cooking

Using dimensional analysis to covert measurements from ounces to cups, cups to pints, pints to quarts, and quarts to gallons

## Disc Golf Connection *(7.RP.1.2.a)(7.NS.2.a)*

Students will follow a *No Bake* recipe to create 'disc' shaped cookies. They will double a recipe for (smaller) putter maker sizes and triple a recipe for (regular) sized putters and drivers.

## Science Connection *(8-9.PS.1.1.1, 1.2.1, 1.2.2)(7.NS.2.d.3)*

Using capacity in dimensional analysis to convert ounces to cups, cups to pints, pints to quarts, and quarts to gallons. Students will also convert from SI units to metrics using grams for weight and liters for liquid volume.

If oz = ounces and g = grams, convert ounces to grams, using the formula: oz x 28.3495 = g.

To convert ounces approximately ≈ to grams, 1 x 30 = g.

## Enrichment *(7.NS.2.c)*

Students will show both *'No Bake'* recipes they created for doubling and tripling their two batches of cookies, converting each from SI standard units to metric units.

## Technology *(6.RP.1.2.3.4.a,b,c,d)(8.EE.5)*

Students will create a Power Point showing how to covert SI standard units to metric units for:

weight (grams)    capacity (liters)    distance (meters)

They will also create a chart in Excel to show equivalency between ounces, cups, pints, quarts, and gallons.

## Anticipatory Math Concepts

Being able to convert temperature from Celsius to Fahrenheit and the U. S. system of yards, feet, and ounces to metric measurements can be important when you're outside on the course playing or when using and buying disc golf supplies here or abroad.

***converting Celsius***- If ***C*** = degrees Celsius and ***F*** = degrees Fahrenheit, the conversion formula from Fahrenheit to Celsius is C = 5/9(F – 32).

***converting Fahrenheit***- To convert Fahrenheit approximately to Celsius, subtract 30 from the Fahrenheit reading and divide the answer in half.

***Celsius to Fahrenheit***- To convert Celsius to Fahrenheit, the formula is F = 1.8 x C + 32.

> <u>*easy approximation*</u>-To convert Celsius approximately to Fahrenheit,
> 2 x C + 30 = F.

## Anticipatory Science Concepts

***metrics to SI***- If ***m*** = meter and ***in.*** = inches, the conversion formula for meters to inches is m x 39.37 = in. To change inches into feet, divide by 12.

**SI to metrics**- To convert inches into centimeters (cm), in. x 2.54 = cm.

***SI to metrics***- If ***mi*** = miles and ***k*** = kilometers, to convert miles to kilometers,

> <u>*easy approximation*</u>- To convert miles approximately to kilometers, mi x 3 divided by 2 = k.

## Anticipatory Disc Golf Concepts

<u>*Conversions while traveling abroad for tournaments*</u>

***metrics to SI***- To convert kilometers to miles, k x 0.6214 = mi.

> <u>*easy approximation*</u>- To convert kilometers approximately to miles,
> k x 2/3 = mi.

***SI to metrics***- If ***oz*** = ounces and ***g*** = grams, to convert ounces to grams, oz x 28.3495 = g.

> <u>*easy approximation*</u>- To convert ounces approximately to grams, 1 x 30 = g.

**SI Units**

| | |
|---|---|
| **8 fluid ounces** | **1 cup** |
| **2 cups** | **1 pint** |
| **2 pints** | **1 quart** |
| **4 quarts** | **1 gallon** |
| **1 yard = 3 feet = 36 inches** | **1 mile = 1760 yards = 5280 feet** |

## Math in Action

**Coach Ramsey had a No Bake Cookie recipe. He thought it would be fun for his disc golf club to make cookies for the coaching staff. The club members received the recipe and knew that they would have to make a large batch to feed all the hungry coaches. They decided to triple the recipe.**

**No Bake Recipe:**

| | |
|---|---|
| **2 cups sugar** | **1/2 cup milk** |
| **1/4 cup butter** | **1/3 cup cocoa** |
| **1/2 cup peanut butter** | **1 tsp vanilla** |
| **3 cups oatmeal** | **1 cup coconut** |

What is known?

The original recipe

What is unknown?

Ingredients needed for a tripled batch

Choose equations to triple the batch:

| | |
|---|---|
| 2(3) = *sugar* | 1/2(3) = *cups milk* |
| 1/4(3) = *butter* | 1/3(3) = *cocoa* |
| 1/2(3) = *peanut butter* | 1(3) = *vanilla* |
| 3(3) = *oatmeal* | 1/2(3) = *coconut* |

*Solution:*

| | |
|---|---|
| *sugar- 6 cups* | *milk- 1.5 cups* |
| *butter- 3/4 cups* | *cocoa- 1 cup* |
| *peanut butter- 1.5 cups* | *vanilla- 3 tsp* |
| *oatmeal- 9 cups* | *coconut- 1.5 cups* |

# LESSON #11

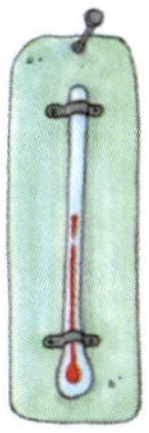

## Math Concept

Using exponential notation

## Disc Golf Connection *(ESS1.k.1)*

Students will keep track of local weather conditions for one week. They will use NWS meteorological symbols to track directional wind speed, precipitation, cloud formations, barometric pressure, and temperature. They will assess whether or not conditions on a specific date are satisfactory or unsatisfactory for safe play.

## Math Connection *(ESS2.D)*

Students will check wind speeds and temperature and use the wind chill index formula to calculate actual wind chill temperatures accurately.

## Science Connection *(ESS2.k.2)*

Using a thermometer, a barometer, and an anemometer, students will create a weather forecast model. They will create a standard weather model diagram using the National Weather Station Symbols.

## Enrichment *(ESS1.k.1)(ESS2.D)(ESS2.k.2)*

Choose an international weather chart online from a local news source and create a forecast model for a major city somewhere on Earth.

## Technology *(ESS.3.B)*

On your Android or iPhone, pick several Apps and follow the weather for several days. Find similarities and differences in local forecasts. Describe and write a short overview of the obvious and not so obvious reasons for distinct variations in weather patterns. Make a graph in Excel to show discrepancies and anomalies in the weather.

[1 Weather, AccuWeather, Carrot Weather, Dark Sky, Google App]

[NOAA Weather, WeatherBug, The Weather Channel]

## Anticipatory Math Concepts

### *Exponential Notation*

***power-*** a product formed from repeated multiplication by the same number

***base-*** the number that is used as a factor in a repeated multiplication

***exponent-*** a number that represents how many times a base is used as a factor in a repeated multiplication

## Anticipatory Science Concepts

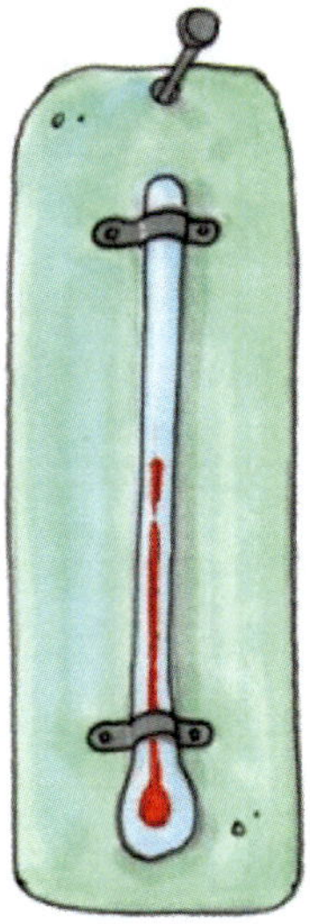

### *Weather Instruments*

***thermometer-*** measures air temperature

***barometer-*** measures air pressure

***sling psychrometer-*** measures relative humidity, with cooling effects of evaporation

***rain gauge-*** measures amount of rain that has fallen

***wind vane-*** determines the direction the wind is blowing

***anemometer-*** measures wind speed

***hygrometer-*** measures the amount of water vapor in the air

***weather satellites-*** track and photograph large-scale air movements

## Anticipatory Disc Golf Concepts

### *Effects of Wind* *drivers **putting

*** head wind-*** disc fades later (faster, more stable, more power)

*** tail wind-*** disc fades sooner and harder (slower, less stable, more fade)

*** cross wind-*** distance (wind lifting disc) accuracy (wind dropping disc) (stay low)

***** head wind-*** (flat nose angle, comfortable wing angle, less glide)

***** tail wind-*** (flat to nose up release, comfortable wing angle, play for a drop)

***** cross wind-*** (flat wing angle, comfortable nose angle, flat release)

# Weather Station Symbols

| | | | | | | | | | |
|---|---|---|---|---|---|---|---|---|---|
| **CLOUD TYPES** | Cirrus | Scattered Cirrus | Dense Cirrus | Cirro Stratus | Heavy Cirro | Cumulus | Medium Cumulus | Cumulo Nimbus | Cirro Cumulus |
| **CLOUD TYPES** | Altotratus Thin | Altostratus Thick | Alto Cumulus Thin | Alto Cumulus Thick | Strato Cumulus | Nimbo Stratus | Stratus | Fracto Stratus | Fracto Stratus Thick |
| **FRONTS** | Warm | Cold | Stationary | Occluded | WIND DIRECTION: N, NE, E, SE, S, SW, W, NW | | | | |
| **CLOUD COVERAGE** | Clear | 1/10 | 1/4 | 1/2 | 3/4 | 9/10 | Overcast | Obscured | Stable |
| **SKY COVER** / **AIR PRESSURE** | Haze | Smoke | Heavy Haze | Fog Patches | Light Fog | Moderate Fog | Heavy Fog | High H | Low L |
| **WIND SPEED** | Calm | < 5 knots | 5 knots | 10 knots | 20 knots | 25 knots | 50 knots | < 75 knots | > 100 knots |
| **STANDARD CONDITIONS** | Rain Light | Rain Moderate | Rain Heavy | Snow Light | Snow Moderate | Snow Heavy | Drizzle Light | Drizzle Moderate | Drizzle Heavy |
| | | | | Weather | Conditions | | | | |
| **STORM FEATURES** | Rain Slight | Rain Moderate | Rain Heavy | Snow Slight | Snow Moderate | Sleet Hail | Freezing Drizzle Light | Freezing Drizzle Heavy | Slick Road Condition |
| **THUNDER STORMS** | Rain Mild | Rain Moderate | Rain Severe | Snow Mild | Snow Moderate | Snow Severe | Hail Mild | Hail Moderate | Hail Severe |
| | | | | **Barometric** | **Tendency** | | | | |
| **UNIQUE** | **FEATURES** | **INCREASE** | **IN** | **AIR** | **PRESSURE** | **DECREASE** | **IN** | **AIR** | **PRESSURE** |
| Hail<br>Snow Grains<br>Tornado<br>Ice Crystals<br>Drifting Snow | | Rising, Then Falling | Rising, then Steady | Rising Steadily | Falling, then Rising | Falling, then Rising | Falling, then Steady | Falling and Steadily | Rising, then Falling |

## Math in Action

**You and a teammate are asked to go play a disc golf tournament in Bari, Italy. The weather will determine whether you can play safely and effectively. The director sends you a model of a weather forecast, which shows the predicted climate for a day late in March. From the weather station model below, would it be advisable to plan your trip overseas?**

What is known? *(ESS.2.D)(ESS.3.A,B,C)(ETS.1.B)*

| | | | |
|---|---|---|---|
| A- | Temperature | F- | Pressure Tendency |
| B- | Present Weather | G- | Wind Speed/Direction |
| C- | Dew Point | H- | Barometric Pressure |
| D- | Low Cloud Type | I- | High Cloud Type |
| E- | Pressure Change | J- | Cloud Coverage |

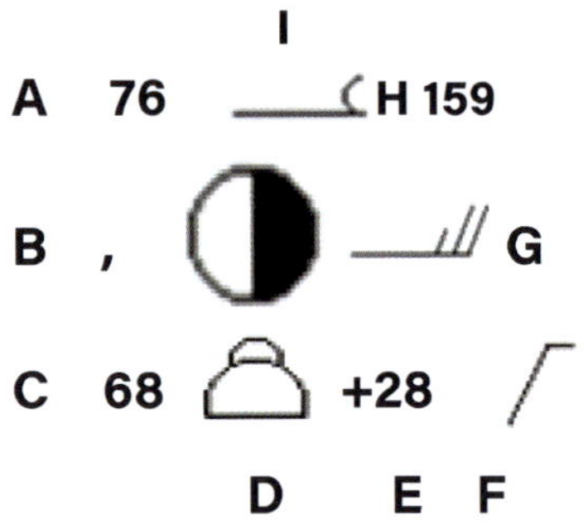


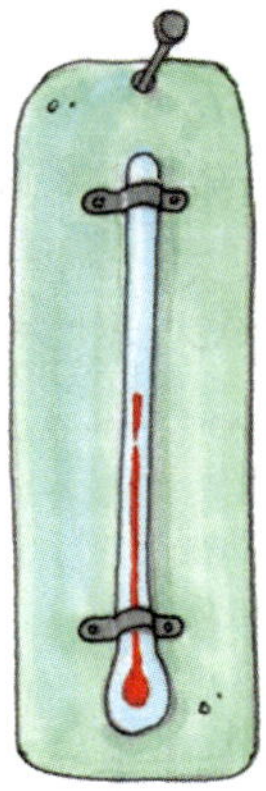

What is unknown?

Barometric pressure is rising from a low (+28). Air rises in a low pressure area. The steady rising air cools and is *likely* to condense water vapor and form clouds, and consequently rain.

Check weather symbols and numbers: 68 - humid , - moist air 159/+28 - unstable air

Solution

With a slight drizzle, cirrostratus clouds mean the winds are showing unstable conditions and a front is stabilizing. A balmy 76° is comfortable, but with wind speeds of 25 knots (≈ 30 mph) you probably should reconsider the offer. If you are brave, take your rain gear!

# Math in Action

Temperature (°F)

Wind (mph)

| CALM | 40 | 35 | 30 | 25 | 20 | 15 | 10 | 5 | 0 | -5 | -10 | -15 | -20 | -25 | -30 | -35 | -40 | -45 |
|---|---|---|---|---|---|---|---|---|---|---|---|---|---|---|---|---|---|---|
| 5 | 36 | 31 | 25 | 19 | 13 | 7 | 1 | -5 | -11 | -16 | -22 | -28 | -34 | -40 | -46 | -52 | -57 | -63 |
| 10 | 34 | 27 | 21 | 15 | 9 | 3 | -4 | -10 | -16 | -22 | -28 | -35 | -41 | -47 | -53 | -59 | -66 | -72 |
| 15 | 32 | 25 | 19 | 13 | 6 | 0 | -7 | -13 | -19 | -26 | -32 | -39 | -45 | -51 | -58 | -64 | -71 | -77 |
| 20 | 30 | 24 | 17 | 11 | 4 | -2 | -9 | -15 | -22 | -29 | -35 | -42 | -48 | -55 | -61 | -68 | -74 | -81 |
| 25 | 29 | 23 | 16 | 9 | 3 | -4 | -11 | -17 | -24 | -31 | -37 | -44 | -51 | -58 | -64 | -71 | -78 | -84 |
| 30 | 28 | 22 | 15 | 8 | 1 | -5 | -12 | -19 | -26 | -33 | -39 | -46 | -53 | -60 | -67 | -73 | -80 | -87 |
| 35 | 28 | 21 | 14 | 7 | 0 | -7 | -14 | -21 | -27 | -34 | -41 | -48 | -55 | -62 | -69 | -76 | -82 | -89 |
| 40 | 27 | 20 | 13 | 6 | -1 | -8 | -15 | -22 | -29 | -36 | -43 | -50 | -57 | -64 | -71 | -78 | -84 | -91 |
| 45 | 26 | 19 | 12 | 5 | -2 | -9 | -16 | -23 | -30 | -37 | -44 | -51 | -58 | -65 | -72 | -79 | -86 | -93 |
| 50 | 26 | 19 | 12 | 4 | -3 | -10 | -17 | -24 | -31 | -38 | -45 | -52 | -60 | -67 | -74 | -81 | -88 | -95 |
| 55 | 25 | 18 | 11 | 4 | -3 | -11 | -18 | -25 | -32 | -39 | -46 | -54 | -61 | -68 | -75 | -82 | -89 | -97 |
| 60 | 25 | 17 | 10 | 3 | -4 | -11 | -19 | -26 | -33 | -40 | -48 | -55 | -62 | -69 | -76 | -84 | -91 | -98 |

**Wind Chill (°F) = $35.74 + 0.6215T - 35.75 (V^{0.16}) + 0.4275T (V^{0.16})$**

**Wind Chill: Frostbite ■ 30 min ■ 10 min ■ 5 min.**

**Green means no real threat. Blue means starting to become dangerous. Purple means life threatening.**

What is known?

wind speed, (v) = 30 mph

temperature, (T) = 25° F

What is unknown?

wind chill

Choose an equation

$$WC = 35.74 + 0.6215T - 35.75 (V^{0.16}) + 0.4275T (V^{0.16})$$

Solution

$35.74.+ 0.6215(25) = 51.278$ $\quad$ $35.75(30^{0.16}) = 61.605$ $\quad$ $51.278 - 61.605 = -10.327$

$0.4275(25)(30^{0.16}) = 18.417$ $\quad$ so, $-10.327 + 18.417 = 8.09°$

**The wind chill is approximately ≈ 8° F** *(ESS.2.D)(ESS.3.A,B,C)(ETS.1.B)*

# LESSON #12

## Math Concept

Using parallels of latitude and longitude to establish direction

## Disc Golf Connection *(ELA-Literacy.RH.6-8.7)*

Find assorted disc golf tournaments played internationally and use latitude and longitude to find directional positioning of exact countries on a political map.

## Math Connection (*ELA-Literacy.RH.6-8.5)*

Students will use a coordinate plane to divide the Earth into topographical regions to establish the locations of exact cities. Using meridians and parallels of latitude, students will identify how lines (Prime Meridian, International Date Line) and circles of demarcation (Arctic Circle, Tropic of Cancer, Equator, Tropic of Capricorn, and the Antarctic Circle) allow us to find and pinpoint exact positions of geographical regions on our planet.

## Science Connection (*ELA-Literaacy.RH.6-8.4)*

Students will determine degrees, minutes, and seconds to establish exact locations.

$$1° = 60' = 3600''$$

For finer accuracy, they will use sexagesimal notation where fractions of seconds are given by a decimal point.

## Enrichment *(ELA-Literacy.RH.6-8.5)(ELA-Literacy.RH.6-8.7)*

Given several latitude and longitude coordinates with their accompanying minutes, degrees, and seconds, students will practice identifying locations of disc golf tournaments across the globe on a political map.

## Technology *(ELA-Literacy.RH.6-8.5)(ELA-Literacy.RH.6-8.7)*

Use the following iPhone apps to practice locating specific landmarks from a given position, then practice navigating to a location of your choice successfully.

[GPS Kit] [MotionX GPS] [Gaia GPS] [iTopo Maps] [Trails - GPS Tracker]

[Sun Compass] [Park Maps] [Accu Terra] [Geocaching] [Elevation Pro]

## Anticipatory Math Concepts

### *Grid Planes*

***coordinate plane-*** a system formed by the intersection of horizontal (x-axis) and vertical (y-axis) number lines

***ordered pairs-*** pairs of numbers or coordinates that can be used to represent points or locations in a coordinate plane, such as found on a political map

## Anticipatory Science Concepts

Converting Degrees to D°M'S"

***degrees-*** use the whole number part of the decimal

***minutes-*** multiply the remaining decimal by 60 and use the whole number part of the answer as minutes

***seconds-*** multiply the new remaining decimal by 60

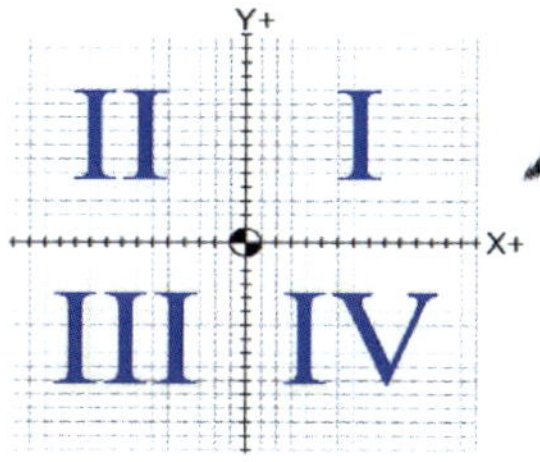


***hyzer angle-*** a disc that is already tilted in the direction that it is going to fade at the end of its shot towards the ground

***anhyzer angle-*** a disc tilted up against the angle it is going to fade at the end of its flight

## Anticipatory Disc Golf Concepts

### *Shot Shapes*

| TYPE | HYZER | FLAT | ANHYZER |
|---|---|---|---|
| OVER STABLE | **Spike Hyzer** short and accurate | **Straight Hyzer** holds angle, skips left, fades out | **Anhyzer Flex** fights out flex to left, fades short |
| STABLE | **Hyzer Flip** hyzer angle will flip to flat or turns, long and straight with some control | **Distance Flex** glides far, holds angle, fades at end, flat for forward penetration | **Max Distance** big anhyzer, holds angle longer, fades at end with more forward penetration |
| UNDER STABLE | **Hyzer Flip** exaggerated hyzer angle, flips to flat, turns over, long flight, fades right | **Anhyzer** big anhyzer, higher, holds angle up and around obstacles, penetrating to right | **Roller** flips over and dives into ground and rolls, curves to left, stands up, goes straight and finishes right- for low branch shot |

# Find the City and Country of Each Tournament

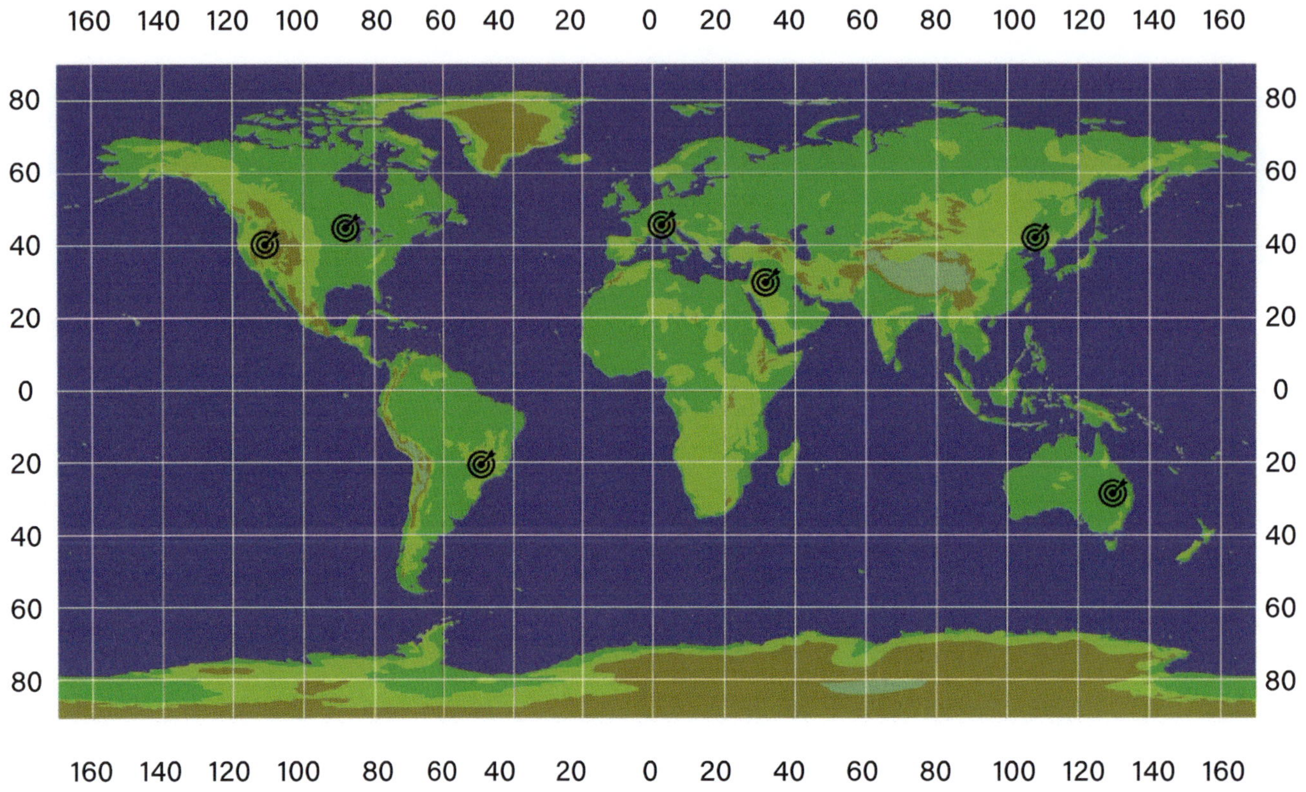


Convert to a decimal: **2° 27′ 47″** = 2° + [27′ · 1°/60′] + [47″ · 1°/3600″] = [2 + 27/60 + 47/3600] ≈ **2.463°**

Convert to D°M′S″: **5.128° = 5°** + [0.128° · 60′/1° = 7.68′] [7.68′ = **7′** + 0.68′] [0.68′ · 60″/1° = **40.8″**]
D°M′S″ ≈ **5°7′41″**

**Convert the decimal coordinates to degrees, minutes, and seconds and identify the city name and country (abbreviation) the tournament is being played in.**

43.6532 N = _____° _____′ _____″ 79.3832 W = _____° _____′ _____″ ________________

33.8688 S = _____° _____′ _____″ 151.2093 E = _____° _____′ _____″ ________________

30.0444 N = _____° _____′ _____″ 31.2357 E = _____° _____′ _____″ ________________

43.6150 N = _____° _____′ _____″ 116.2023 W = _____° _____′ _____″ ________________

48.8566 N = _____° _____′ _____″ 2.3522 E = _____° _____′ _____″ ________________

22.9068 S = _____° _____′ _____″ 43.1729 W = _____° _____′ _____″ ________________

39.9042 N = _____° _____′ _____″ 116.4072 E = _____° _____′ _____″ ________________

# LESSON #13

## Math Concept

Fractional Proportionality

## Disc Golf Connection *(MU.Cr1.1.6-1.1.8a, 2.1.6-2.1.8a, 3.1.6-3.1.8a)*

Using disc-shaped putter markers to show notes while reading musical notation in the bass and treble clefs

## Math Connection *(6.RP.3.a,b,d)(7.RP.1.2.a,b)(8.EE.3,4)*

Reading time signatures to see how fractional relationships affect variations in musical notes through rhythm - (4/4): 1 whole note = 2 half notes = 4 quarter notes = 8 eighth notes = 16 sixteenth notes

## Science Connection *(6.EE.1-6)(7.EE.4)(8.EE.1-4)(PS.4.MS.1,2,3)(PS.4.A,B,C)*

Showing wave frequencies to analyze the pitches of sound in a standard medium. Students will study vibration, resonance, and frequency in musical instruments. They will analyze how pitch is controlled in wind instruments by closing and opening specific valves. They will examine the length of the air column, the wavelength, and how the frequencies are changed to affect pitch.

Wrapping paper roll experiment with rubber bands

Frequency = velocity/wavelength 34,400 cm/s $(3.44 \times 10^4)$ cm/s

## Enrichment *(6.RP.3.a,b,c,d)(MU.Cr.2.1.6-2.1.8b)*

Students will use 20 small putter markers with 10 colored pieces of yarn for staff lines respectively- (5 blue for bass clef) and (5 red for treble clef).

They will place each marker on and between the staff lines to represent each note on the bass and treble clefs. After placement, they will take a black visa-vi pen and write the correct names of each note in sequential order, then scramble and reassemble in the proper order. Yarn may also be used to attach the correct positions of the stem lines. (third line and above- stem on left side facing down) (below third line- stem on right side facing up)

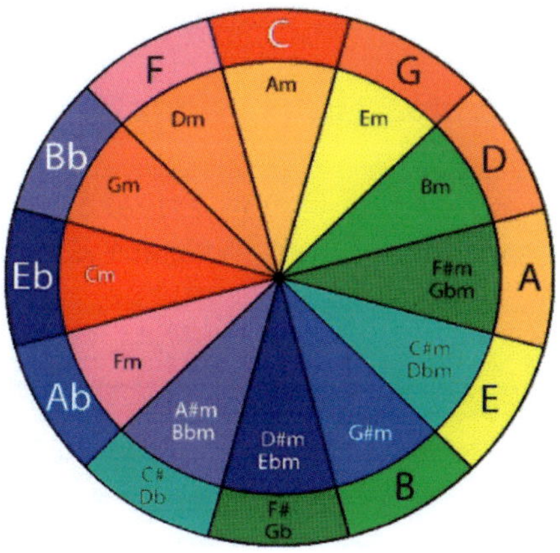


## Technology *(6.RP.3.c)(7.RP.1,2.a.b.c)(7.NS.1.d)*

Creating a pie graph to display the

Circle of $5^{th}$ Chords data on an Excel Spreadsheet.

## Anticipatory Math Concepts

### *Proportionality*

***proportion-*** having the same or a constant ratio (measured by division) or relation

***rate-*** a ratio of two related quantities that are measured in varying units

***rational number-*** a number that can be written as a fraction in a/b form

## Anticipatory Science Concepts

### *Sound*

***pitch-*** frequencies of sounds waves that determine high and low sounds

***frequency-*** wave crests that will pass a certain point every second, shown as hertz (Hz)

***amplitude-*** distances in a wave from the rest position of the medium to the trough or crest position

***wavelength-*** a point on one wave and the distance to the same point on the proceeding wave

***waves-*** rhythmic anomalies that carry kinetic energy transfers through matter or space

***crest-*** the highest spot or point on a wave

***trough-*** the lowest spot or point on a wave

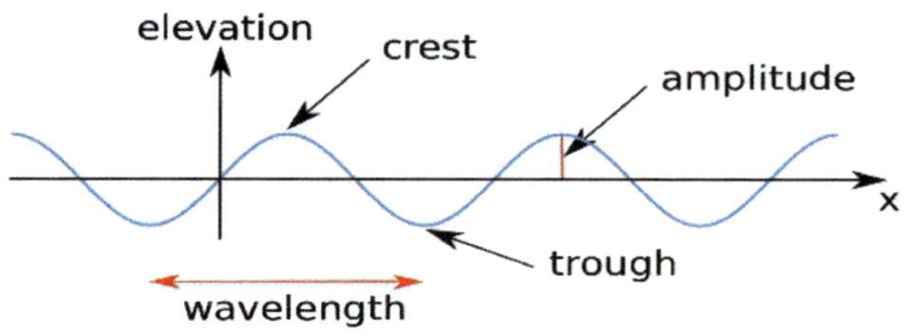


## Anticipatory Disc Golf Concepts

***speed-*** How a disc's momentum carries it through the air. Higher numbered discs fly faster, and require more power to gain fluency and upward flight.

***glide-*** The disc's design to stay in flight while gaining speed. Discs that have more glide tend to pull in a downward direction during flight time. Beginning players in this sport should look for discs rated a number 6 that have the most glide.

***turn-*** The tendency of discs to bank to the right (for RHBH throwers) in early flight. Some discs resist turning and are rated at a +1. Beginning players should use discs that turn easily and are rated at a -5.

***fade-*** All discs tend to trail off in a definite direction, depending on which hand you use to throw with. Discs often hook or will remain in a constant direction. The straightest finishing pattern for a disc is rated 0, while a disc that will hook is rated at a 5.

# Math in Action

**Mr. Walstrom, the music teacher and disc golf coach, wants to give a note value quiz in music class, but he realized he needs the entire class time to practice for the spring festival.**

**He decides to have the students take the quiz outside on the course after school during club. He gives the students cards with the pictures of whole notes, half notes, quarter notes, eighth notes and sixteenth notes.**

**He asks the students to create a 3-D visual diagram of the breakdown of whole notes to sixteenth notes in a 4/4 signature, using a basket full of yellow, green, red, blue, and purple putter markers. He said the markers must show the entire diagram representing the values of each type of note in the correct order of ratio proportionality. (7.RP.A.2, 7.RP.A.2c)**

What is known?

> the types of notes to be used and the time signature

What is unknown?

> the correct order to place the notes in proportional order

Choose an equation to show equal ratios between the notes

> **$k = y/x$ half note**

**Solution** *7.RP.A.2, 7.RP.A.2c*

$.5 = 2/4$ $k(x) = y$ $.5(4) = 2$

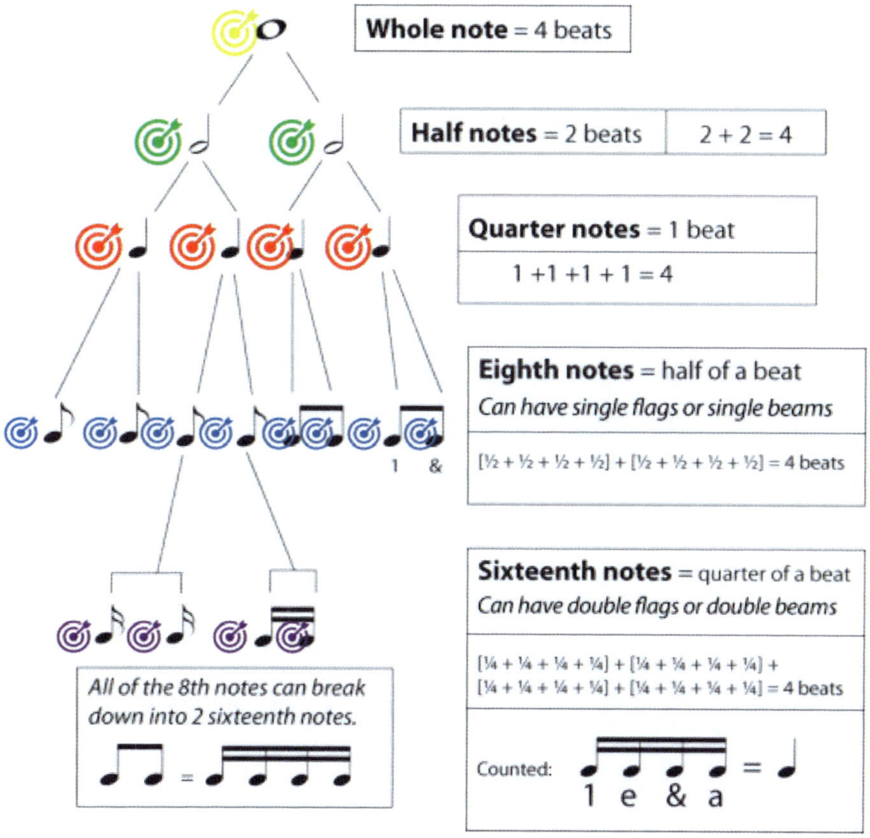

# GLOSSARY

## Math

***acute***- An angle smaller than 90º.

***area***- The interior measure of a circle, as used in the formula $\pi r^2$

***average rate***- Dividing the distance traveled by the travel time

***circumference-*** The distance around a circle, used in the formula $2\pi d$

***diameter***- Connecting points on the perimeter of a circle that pass through the center

***frequency of an interval***- The amount of numbers appearing in an interval

***frequency table-*** Used in grouping intervals of numerical values

***gridlines***- Used to show variations in numerical values of the data displayed

***histogram***- A graph that shows numerical data in a frequency table. Individual bars are used to show each numerical interval. Frequency is used by showing varying bar heights

***hypotenuse***- The side of a triangle that is exactly opposite to the right angle

***Legend-*** *D*escribes types of data used on the graph

***legends***- Describe types of data used on the graph

***legs***- In a right triangle, the sides that form the right angle

***mean***- The sum of a set of numbers divided by the amount of numbers in the group

***median***- The middle or center number when a group is written in numerical order. When the data set has an even number of numbers, the mean is the average of the 2 middle numbers.

***mode-*** The number that occurs most often. Some sets have no mode, or can have more than 1.

***obtuse-*** An angle larger than 90º

***Pythagorean Theorem***- For all right triangles, adding the sum of the squares of the lengths of the legs is equal to the square of the length of the hypotenuse

$(a^2 + b^2 = c^2)$

***radius***- Any point on the perimeter of a circle and its distance to the center

***range***- Subtract or find the difference between the smallest and largest number

***rate-*** A ratio of two related quantities measured in varying units (unit rate has a denominator of one)

***scales***- Increments that show lengths of bars and lines numerically

***scatter plot***- Numerical pairs on a graph shown as (x,y), which plots collective points in a coordinate plane

***stem-and-leaf-*** Arranging numerical data in a way that shows and demonstrates mean, median, mode, and range in a graphing format

## Science

***acceleration***- The change of velocity at which an object is traveling

***center of mass***- The point which acts as though the total mass of the object were at that position

***centrifugal force-*** The equal and opposite reaction to the centripetal force. It does not act on the object moving in the circle and is not considered.

***centripetal acceleration***- The acceleration of an object in circular motion acting towards the center of the circle. $A = v^2/r$ v = velocity around the circle r = radius of circle

***centripetal force-*** The force which acts on an object towards the center of a circle to produce centripetal acceleration, and so keeps the object moving in a circle

***efficiency***- The work done on the load divided by the work done by the effort, expressed as a percentage

***effort arm***- A point on a lever where the effort force is applied

***escape velocity-*** The smallest speed at which a body will travel in order to escape the gravitational tug of another body without any extended force

***first class lever-*** Where the fulcrum is located between the effort and resistance forces

***free fall-*** The unrestricted motion of an object when it is acted upon only by the single gravitational force when there are no resistive or other forces acting, like air resistance

***fulcrum-*** The part of a lever that has a fixed point

***gram-*** Used to measure mass or weight in the metric system

***inertia***- Whether or not an object will resist a change in motion

***lever-*** A free pivoting bar that can turn, attached to a definite fixed point

***linear motion***: Movement in a straight line

***liter***- Used to measure capacity in the metric system

***machine***- A device that is used to supersede a force or (load)

***mass-*** The quantity of matter within an object

***meter***- Used to measure length in the metric system

***momentum***- The mass of an object multiplied by its velocity

***motion-*** The change in position of a moving object

***Newton's Laws-*** (1) If an object rests, or has a consistency in speed or direction, the force on it is zero.

(2) When the momentum of an object accelerates, then there is a separate force acting on it. Objects usually have a constant force that is proportional to their acceleration. The force and acceleration move in the same direction.

(3) Forces occur in equal and opposite pairs called the action and reaction. When said object A exerts a force on object B, object

B exerts an equal and opposite force on A. These forces are balanced.

***Newton's Law of Gravitation-*** States that there is a gravitational force of attraction between any two objects with mass, which depends on their masses and the distance between them.

***resistance arm***- The point on a lever where the resistance force is exerted

***second class lever***- When a resistance force is located between the effort force and the fulcrum that always multiplies force

***speed-*** The distance an object moves in a specific time interval

***terminal velocity-*** The largest, constant speed reached by a falling body passing through a gas or liquid. As the speed increases, the resistance due to air or liquid (drag) increases. Eventually, the weight will equalize or balance out.

***third class lever***- When an effort force is located between the fulcrum and the resistance force. The effort arm is shorter than the resistance arm, so it can't multiply force and the mechanical advantage (MA) is always less than one.

***velocity***- The direction or speed an object is traveling

## Disc Golf

***courtesy***- Players should not throw until they are certain the thrown disc will not distract another player or injure anyone present. Players should take care to not distract other players while it is their turn. Littering on the course is discouraged and considered a courtesy violation. Players are expected to watch where other players' discs go and search for discs in the event they are lost.

***discs-*** Today discs are much smaller and heavier than traditional flying discs. They measure 8-9 inches (20-23cm) in diameter and have a mass of 120-180 grams.

***discs used in play***- Playing discs must adhere to the conditions put forth by the PDGA Technical Standards. Any player that uses illegal discs will receive two penalty throws.

***drivers***- These discs have sharp angles edges and carry most of their mass towards the outer rim of the disc. The driver was created to travel longer distances at greater speeds of acceleration.

***initiating position-*** After a disc is thrown, the exact area where it first comes to rest, when the disc has completely stopped all motion.

***fade-*** The slowing down of the disc nearing the end of its flight.

***features***- The design of modern discs are shaped for speed, precision, and control. They are categorized as: putters, mid-range discs, and drivers.

***holing out***- When holing out on a basket target, the disc must come to rest within the bottom cylinder of the basket or inside of the chains.

***maintaining the lie***- The original position where the disc lands after being thrown. A small mini disc marker is generally placed directly behind where the disc hit the ground coming to a resting position.

***mid-range discs***- Disc that have sharper angles edges that allow them to cut through the air more easily. They generally travel faster, have more stability, and will fly further than a putter.

***out of bounds***- When a disc crosses into a place that exceeds the perimeter of the designated area of play, penalty throws are charged and can be thrown up to a meter from the previous lie and perpendicular to the fowl position.

***overhand putt***- When a player holds a disc level, supported in the center by the thumb, with the finger tips beyond the edge, as if you were supporting a plate. The player extends his opposite leg away from the throwing arm, reaches back and then points their arm towards the basket, tossing the disc in a fashion similar to when throwing a dart. Ideally the thrower wouldn't turn or rotate the wrist. A gentle follow through will generally give the disc its spin. Sometimes lifting the opposite leg will ensure a proper release.

***over-stable***- Disc's flight moves to the left (for a right-handed, backhand throw).

***putters-*** Discs created to fly straight, accurately, and predictably at slow rates of speed. They conform well with controlled shots that are near the basket and are used for short drives when shrubs, wooded areas or other obstacles are in the player's range.

***rotation-*** The spin doesn't affect the lift and drag forces much, but determines a disc's stability while in flight.

***signage***- These are important in any and all disc golf course designs. They show accurate distances, boundary lines, par counts, and the layout of the entire course. They often identify mandatory paths, boundary lines, and lengths for the players to see clearly while the play.

***spin rate-*** A disc's spin rate does not change while it loses velocity depending on its sloped design angle.

***Stability-*** Measuring a disc's ability to move side to side during its flight.

***targets***- The Disc Pole Hole is often called the basket for general courses in disc golf.

***tee pads***- A starting pad is usually made of concrete and signifies where a player begins the hole. Mulch, decomposed granite, wood, or other natural materials are often found at the pads today.

***teeing off-*** Players begin at the tee box. A player's foot must be touching in the tee box at the time they throw the disc.

***throwing from a stance***- While releasing a disc, a player must have one foot in direct contact with the playing surface on the lie.

***topography***- Disc golf is unique in that it utilized Mother Nature's design! Trees, shrubs, wooded areas, elevation changers, creeks and ponds make each course unique, challenging, and fun.

***turn***- The disc's turning pattern from the start to the middle of its flight.

***underhand putt-*** The player pushes the disc towards the basket from the solar plexus. Upon release of the disc, continue to point with your index finger towards the target. The more wrist you include, the greater the spin. The thumb should be pointing straight up, as though shaking someone's hand.

***under-stable-*** Disc's flight moves to the right (for a right-handed, backhand throw).

***using courtesy***- Players should and will not throw until they are cognizant that the thrown disc will not distract other players or injure anyone present. Littering on any disc golf course is discouraged and considered a courtesy violation. Players are encouraged to watch where other players discs land and search for lost discs when possible and plausible.

## Enrichment

***Absolute Value***

(Add/Subtract Integers)

Using integers to describe scoring in Disc Golf. Let students cut out their own footprints, then use them as place holders to add and subtract integers as they move the prints up and down the number line on the floor.

***Central Angles***

(Name and Identify)

Radius, diameter, central angle, and chord

***Constants of Proportionality***

(Percentages/ Angles)

Using bar graphs to find constants of proportionality in percentages using means/ extremes [k(x) = y] Identify acute and obtuse angles.

***Integers***

(Scoring)

Describe how to score in disc golf.

'+' positives/ '-' negatives

For integers with different signs, subtract the lesser absolute value from the greater absolute value. Use the sign of the number with the greater absolute value.

'+' positives/ '-' negatives

For integers with the same sign, add the absolute values and use the common sign.

***Mechanical advantage***

(Levers)

Explain and identify how the human arm is used as a 3rd class lever.

***Metric Systems*** (Conversions)

Using a metric ladder, show conversions in the metric system for weight (grams), capacity (liters), and distance (meters).

***Rates and Unit Rates***

(Using Rates for Data)

Calculate average rate as distance divided by time.

# Idaho CCS Math Standards and Essentials

## GRADE 6

*Ratios and Proportional Relationships - 6.RP*

### Understand ratio concepts and use ratio reasoning to solve problems.

1. Understand the concept of a ratio and use ratio language to describe a ratio relationship between two quantities. For example, "The ratio of wings to beaks in the bird house at the zoo was 2:1, because for every 2 wings there was 1 beak." "For every vote candidate A received, candidate C received nearly three votes."
2. Understand the concept of a unit rate a/b associated with a ratio a:b with b ≠ 0, and use rate language in the context of a ratio relationship. For example, "This recipe has a ratio of 3 cups of flour to 4 cups of sugar, so there is 3/4 cup of flour for each cup of sugar." "We paid $75 for 15 hamburgers, which is a rate of $5 per hamburger."1
3. Use ratio and rate reasoning to solve real-world and mathematical problems, e.g., by reasoning about tables of equivalent ratios, tape diagrams, double number line diagrams, or equations.
   a. Make tables of equivalent ratios relating quantities with whole- number measurements, find missing values in the tables, and plot the pairs of values on the coordinate plane. Use tables to compare ratios.
   b. Solve unit rate problems including those involving unit pricing and constant speed. For example, if it took 7 hours to mow 4 lawns, then at that rate, how many lawns could be mowed in 35 hours? At what rate were lawns being mowed?
   c. Find a percent of a quantity as a rate per 100 (e.g., 30% of a quantity means 30/100 times the quantity); solve problems involving finding the whole, given a part and the percent.
   d. Use ratio reasoning to convert measurement units; manipulate and transform units appropriately when multiplying or dividing quantities.

*The Number System - 6.NS*

### Apply and extend previous understandings of multiplication and division to divide fractions by fractions.

1. Interpret and compute quotients of fractions, and solve word problems involving division of fractions by fractions, e.g., by using visual fraction models and equations to represent the problem. For example, create a story context for (2/3) ÷ (3/4) and use a visual fraction model

to show the quotient; use the relationship between multiplication and division to explain that (2/3) ÷ (3/4) = 8/9 because 3/4 of 8/9 is 2/3. (In general, (a/b) ÷ (c/d) = ad/bc.) How much chocolate will each person get if 3 people share 1/2 lb of chocolate equally? How many 3/4-cup servings are in 2/3 of a cup of yogurt? How wide is a rectangular strip of land with length 3/4 mi and area 1/2 square mi?

**Compute fluently with multi-digit numbers and find common factors and multiples.**

2. Fluently divide multi-digit numbers using the standard algorithm.
3. Fluently add, subtract, multiply, and divide multi-digit decimals using the standard algorithm for each operation.
4. Find the greatest common factor of two whole numbers less than or equal to 100 and the least common multiple of two whole numbers less than or equal to 12. Use the distributive property to express a sum of two whole numbers 1–100 with a common factor as a multiple of a sum of two whole numbers with no common factor. For example, express 36 + 8 as 4 (9 + 2).

**Apply and extend previous understandings of numbers to the system of rational numbers.**

5. Understand that positive and negative numbers are used together to describe quantities having opposite directions or values (e.g., temperature above/below zero, elevation above/below sea level, credits/debits, positive/negative electric charge); use positive and negative numbers to represent quantities in real-world contexts, explaining the meaning of 0 in each situation.
6. Understand a rational number as a point on the number line. Extend number line diagrams and coordinate axes familiar from previous grades to represent points on the line and in the plane with negative number coordinates.
   a. Recognize opposite signs of numbers as indicating locations on opposite sides of 0 on the number line; recognize that the opposite of the opposite of a number is the number itself, e.g., -(–3) = 3, and that 0 is its own opposite.
   b. Understand signs of numbers in ordered pairs as indicating locations in quadrants of the coordinate plane; recognize that when two ordered pairs differ only by signs, the locations of the points are related by reflections across one or both axes.
   c. Find and position integers and other rational numbers on a horizontal or vertical number line diagram; find and position pairs of integers and other rational numbers on a coordinate plane.

7. Understand ordering and absolute value of rational numbers.

   a. Interpret statements of inequality as statements about the relative position of two numbers on a number line diagram. For example, interpret –3 > –7 as a statement that –3 is located to the right of –7 on a number line oriented from left to right.

   b. Write, interpret, and explain statements of order for rational numbers in real-world contexts. For example, write –3° C > –7o C to express the fact that –3° C is warmer than –7° C.

   c. Understand the absolute value of a rational number as its distance from 0 on the number line; interpret absolute value as magnitude for a positive or negative quantity in a real-world situation. For example, for an account balance of –30 dollars, write |–30| = 30 to describe the size of the debt in dollars.

   d. Distinguish comparisons of absolute value from statements about order. For example, recognize that an account balance less than –30 dollars represents a debt greater than 30 dollars.

8. Solve real-world and mathematical problems by graphing points in all four quadrants of the coordinate plane. Include use of coordinates and absolute value to find distances between points with the same first coordinate or the same second coordinate.

*Expressions and Equations – 6.EE*

**Apply and extend previous understandings of arithmetic to algebraic expressions.**

1. Write and evaluate numerical expressions involving whole-number exponents.

2. Write, read, and evaluate expressions in which letters stand for numbers.

   a. Write expressions that record operations with numbers and with letters standing for numbers. For example, express the calculation "Subtract y from 5" as 5 – y.

   b. Identify parts of an expression using mathematical terms (sum, term, product, factor, quotient, coefficient); view one or more parts of an expression as a single entity. For example, describe the expression 2 (8 + 7) as a product of two factors; view (8 + 7) as both a single entity and a sum of two terms.

   c. Evaluate expressions at specific values of their variables. Include expressions that arise from formulas used in real-world problems. Perform arithmetic operations, including those involving whole- number exponents, in the conventional order when there are no parentheses to specify a particular order (Order of Operations). For example, use the formulas V = s3 and A = 6 s2 to find the volume and surface area of a cube with sides of length s = 1/2.

3. Apply the properties of operations to generate equivalent expressions. For example, apply the distributive property to the expression $3(2 + x)$ to produce the equivalent expression $6 + 3x$; apply the distributive property to the expression $24x + 18y$ to produce the equivalent expression $6(4x + 3y)$; apply properties of operations to $y + y + y$ to produce the equivalent expression $3y$.

4. Identify when two expressions are equivalent (i.e., when the two expressions name the same number regardless of which value is substituted into them). For example, the expressions $y + y + y$ and $3y$ are equivalent because they name the same number regardless of which number $y$ stands for.

**Reason about and solve one-variable equations and inequalities.**

5. Understand solving an equation or inequality as a process of answering a question: which values from a specified set, if any, make the equation or inequality true? Use substitution to determine whether a given number in a specified set makes an equation or inequality true.

6. Use variables to represent numbers and write expressions when solving a real-world or mathematical problem; understand that a variable can represent an unknown number, or, depending on the purpose at hand, any number in a specified set.

7. Solve real-world and mathematical problems by writing and solving equations of the form $x + p = q$ and $px = q$ for cases in which $p$, $q$ and $x$ are all nonnegative rational numbers.

8. Write an inequality of the form $x > c$ or $x < c$ to represent a constraint or condition in a real-world or mathematical problem. Recognize that inequalities of the form $x > c$ or $x < c$ have infinitely many solutions; represent solutions of such inequalities on number line diagrams.

**Represent and analyze quantitative relationships between dependent and independent variables.**

9. Use variables to represent two quantities in a real-world problem that change in relationship to one another; write an equation to express one quantity, thought of as the dependent variable, in terms of the other quantity, thought of as the independent variable. Analyze the relationship between the dependent and independent variables using graphs and tables, and relate these to the equation. For example, in a problem involving motion at constant speed, list and graph ordered pairs of distances and times, and write the equation $d = 65t$ to represent the relationship between distance and time.

*Geometry – 6.G*

**Solve real-world and mathematical problems involving area, surface area, and volume.**

1. Find the area of right triangles, other triangles, special quadrilaterals, and polygons by composing into rectangles or decomposing into triangles and other shapes; apply these techniques in the context of solving real-world and mathematical problems.

2. Find the volume of a right rectangular prism with fractional edge lengths by packing it with unit cubes of the appropriate unit fraction edge lengths, and show that the volume is the same as would be found by multiplying the edge lengths of the prism. Apply the formulas $V = l\ w\ h$ and $V = b\ h$ to find volumes of right rectangular prisms with fractional edge lengths in the context of solving real-world and mathematical problems.

3. Draw polygons in the coordinate plane given coordinates for the vertices; use coordinates to find the length of a side joining points with the same first coordinate or the same second coordinate. Apply these techniques in the context of solving real-world and mathematical problems

4. Represent three-dimensional figures using nets made up of rectangles and triangles, and use the nets to find the surface area of these figures. Apply these techniques in the context of solving real-world and mathematical problems.

*Statistics and Probability – 6.SP*

### Develop understanding of statistical variability.

1. Recognize a statistical question as one that anticipates variability in the data related to the question and accounts for it in the answers. For example, "How old am I?" is not a statistical question, but "How old are the students in my school?" is a statistical question because one anticipates variability in students' ages.

2. Understand that a set of data collected to answer a statistical question has a distribution which can be described by its center, spread, and overall shape.

3. Recognize that a measure of center for a numerical data set summarizes all of its values with a single number, while a measure of variation describes how its values vary with a single number.

### Summarize and describe distributions.

4. Display numerical data in plots on a number line, including dot plots, histograms, and box plots.

5. Summarize numerical data sets in relation to their context, such as by:

    a. Reporting the number of observations.

    b. Describing the nature of the attribute under investigation, including how it was measured and its units of measurement.

    c. Giving quantitative measures of center (median and/or mean) and variability (interquartile range and/or mean absolute deviation), as well as describing any overall pattern and any striking deviations from the overall pattern with reference to the context in which the data were gathered.

    d. Relating the choice of measures of center and variability to the shape of the data distribution and the context in which the data were gathered.

## GRADE 7

*Ratios and Proportional Relationships – 7. RP*

**Analyze proportional relationships and use them to solve real-world and mathematical problems.**

1. Compute unit rates associated with ratios of fractions, including ratios of lengths, areas and other quantities measured in like or different units. For example, if a person walks 1/2 mile in each 1/4 hour, compute the unit rate as the complex fraction 1/2/1/4 miles per hour, equivalently 2 miles per hour.
2. Recognize and represent proportional relationships between quantities.
   - **a.** Decide whether two quantities are in a proportional relationship, e.g., by testing for equivalent ratios in a table or graphing on a coordinate plane and observing whether the graph is a straight line through the origin.
   - **b.** Identify the constant of proportionality (unit rate) in tables, graphs, equations, diagrams, and verbal descriptions of proportional relationships.
   - **c.** Represent proportional relationships by equations. For example, if total cost t is proportional to the number n of items purchased at a constant price p, the relationship between the total cost and the number of items can be expressed as t = pn.
   - **d.** Explain what a point (x, y) on the graph of a proportional relationship means in terms of the situation, with special attention to the points (0, 0) and r is the unit rate.
3. Use proportional relationships to solve multistep ratio and percent problems. Examples: simple interest, tax, markups and markdowns, gratuities and commissions, fees, percent increase and decrease, percent error.

*The Number System – 7.NS*

**Apply and extend previous understandings of operations with fractions to add, subtract, multiply, and divide rational numbers.**

1. Apply and extend previous understandings of addition and subtraction to add and subtract rational numbers; represent addition and subtraction on a horizontal or vertical number line diagram.
   - **a.** Describe situations in which opposite quantities combine to make 0. For example, a hydrogen atom has 0 charge because its two constituents are oppositely charged.
   - **b.** Understand p + q as the number located a distance |q| from p, in the positive or negative direction depending on whether q is positive or negative. Show that a number and its

opposite have a sum of 0 (are additive inverses). Interpret sums of rational numbers by describing real-world contexts.

   c. Understand subtraction of rational numbers as adding the additive inverse, p – q = p + (–q). Show that the distance between two rational numbers on the number line is the absolute value of their difference, and apply this principle in real-world contexts.

   d. Apply properties of operations as strategies to add and subtract rational numbers.1

2. Apply and extend previous understandings of multiplication and division and of fractions to multiply and divide rational numbers.

   a. Understand that multiplication is extended from fractions to rational numbers by requiring that operations continue to satisfy the properties of operations, particularly the distributive property, leading to products such as (–1)(–1) = 1 and the rules for multiplying signed numbers. Interpret products of rational numbers by describing real-world contexts.

   b. Understand that integers can be divided, provided that the divisor is not zero, and every quotient of integers (with non-zero divisor) is a rational number. If p and q are integers, then –(p/q) = (–p)/q = p/(–q). Interpret quotients of rational numbers by describing real-world contexts.

   c. Apply properties of operations as strategies to multiply and divide rational numbers.

   d. Convert a rational number to a decimal using long division; know that the decimal form of a rational number terminates in 0s or eventually repeats.

3. Solve real-world and mathematical problems involving the four operations with rational numbers.

*Expressions and Equations – 7.EE*

**Use properties of operations to generate equivalent expressions.**

1. Apply properties of operations as strategies to add, subtract, factor, and expand linear expressions with rational coefficients.

2. Understand that rewriting an expression in different forms in a problem context can shed light on the problem and how the quantities in it are related. For example, a + 0.05a = 1.05a means that "increase by 5%" is the same as "multiply by 1.05."

**Solve real-life and mathematical problems using numerical and algebraic expressions and equations.**

1. Solve multi-step real-life and mathematical problems posed with positive and negative rational numbers in any form (whole numbers, fractions, and decimals), using tools strategically. Apply properties of operations to calculate with numbers in any form; convert

between forms as appropriate; and assess the reasonableness of answers using mental computation and estimation strategies. For example: If a woman making $25 an hour gets a 10% raise, she will make an additional 1/10 of her salary an hour, or $2.50, for a new salary of $27.50. If you want to place a towel bar 9 3/4 inches long in the center of a door that is 27 1/2 inches wide, you will need to place the bar about 9 inches from each edge; this estimate can be used as a check on the exact computation.

2. Use variables to represent quantities in a real-world or mathematical problem, and construct simple equations and inequalities to solve problems by reasoning about the quantities.

    a. Solve word problems leading to equations of the form $px + q = r$ and $p(x + q) = r$, where p, q, and r are specific rational numbers. Solve equations of these forms fluently. Compare an algebraic solution to an arithmetic solution, identifying the sequence of the operations used in each approach. For example, the perimeter of a rectangle is 54 cm. Its length is 6 cm. What is its width?

    b. Solve word problems leading to inequalities of the form $px + q > r$ or $px + q < r$, where p, q, and r are specific rational numbers. Graph the solution set of the inequality and interpret it in the context of the problem. For example: As a salesperson, you are paid $50 per week plus $3 per sale. This week you want your pay to be at least $100. Write an inequality for the number of sales you need to make, and describe the solutions.

*Geometry – 7.G*

**Draw, construct, and describe geometrical figures and describe the relationships between them.**

1. Solve problems involving scale drawings of geometric figures, including computing actual lengths and areas from a scale drawing and reproducing a scale drawing at a different scale.

2. Draw (freehand, with ruler and protractor, and with technology) geometric shapes with given conditions. Focus on constructing triangles from three measures of angles or sides, noticing when the conditions determine a unique triangle, more than one triangle, or no triangle.

3. Describe the two-dimensional figures that result from slicing three- dimensional figures, as in plane sections of right rectangular prisms and right rectangular pyramids.

**Solve real-life and mathematical problems involving angle measure, area, surface area, and volume.**

4. Know the formulas for the area and circumference of a circle and use them to solve problems; give an informal derivation of the relationship between the circumference and area of a circle.

5. Use facts about supplementary, complementary, vertical, and adjacent angles in a multi-step problem to write and solve simple equations for an unknown angle in a figure.

6. Solve real-world and mathematical problems involving area, volume and surface area of two- and three-dimensional objects composed of triangles, quadrilaterals, polygons, cubes, and right prisms.

*Statistics and Probability 0 7.SP*

**Use random sampling to draw inferences about a population.**

1. Understand that statistics can be used to gain information about a population by examining a sample of the population; generalizations about a population from a sample are valid only if the sample is representative of that population. Understand that random sampling tends to produce representative samples and support valid inferences.
2. Use data from a random sample to draw inferences about a population with an unknown characteristic of interest. Generate multiple samples (or simulated samples) of the same size to gauge the variation in estimates or predictions. For example, estimate the mean word length in a book by randomly sampling words from the book; predict the winner of a school election based on randomly sampled survey data. Gauge how far off the estimate or prediction might be.

**Draw informal comparative inferences about two populations.**

3. Informally assess the degree of visual overlap of two numerical data distributions with similar variabilities, measuring the difference between the centers by expressing it as a multiple of a measure of variability. For example, the mean height of players on the basketball team is 10 cm greater than the mean height of players on the soccer team, about twice the variability (mean absolute deviation) on either team; on a dot plot, the separation between the two distributions of heights is noticeable.
4. Use measures of center and measures of variability for numerical data from random samples to draw informal comparative inferences about two populations. For example, decide whether the words in a chapter of a seventh-grade science book are generally longer than the words in a chapter of a fourth-grade science book.

**Investigate chance processes and develop, use, and evaluate probability models.**

5. Understand that the probability of a chance event is a number between 0 and 1 that expresses the likelihood of the event occurring. Larger numbers indicate greater likelihood. A probability near 0 indicates an unlikely event, a probability around 1/2 indicates an event that is neither unlikely nor likely, and a probability near 1 indicates a likely event.
6. Approximate the probability of a chance event by collecting data on the chance process that produces it and observing its long-run relative frequency, and predict the approximate relative

frequency given the probability. For example, when rolling a number cube 600 times, predict that a 3 or 6 would be rolled roughly 200 times, but probably not exactly 200 times.

7. Develop a probability model and use it to find probabilities of events. Compare probabilities from a model to observed frequencies; if the agreement is not good, explain possible sources of the discrepancy.
   - **a.** Develop a uniform probability model by assigning equal probability to all outcomes, and use the model to determine probabilities of events. For example, if a student is selected at random from a class, find the probability that Jane will be selected and the probability that a girl will be selected.
   - **b.** Develop a probability model (which may not be uniform) by observing frequencies in data generated from a chance process. For example, find the approximate probability that a spinning penny will land heads up or that a tossed paper cup will land open-end down. Do the outcomes for the spinning penny appear to be equally likely based on the observed frequencies?
8. Find probabilities of compound events using organized lists, tables, tree diagrams and simulation.
   - **a.** Understand that, just as with simple events, the probability of a compound event is the fraction of outcomes in the sample space for which the compound event occurs.
   - **b.** Represent sample spaces for compound events using methods such as organized lists, tables and tree diagrams. For an event described in everyday language (e.g., "rolling double sixes"), identify the outcomes in the sample space which compose the event.
   - **c.** Design and use a simulation to generate frequencies for compound events. For example, use random digits as a simulation tool to approximate the answer to the question: If 40% of donors have type A blood, what is the probability that it will take at least 4 donors to find one with type A blood?

## GRADE 8

*The Number System – 8.NF*

**Know that there are numbers that are not rational, and approximate them by rational numbers.**

1. Know that numbers that are not rational are called irrational. Understand informally that every number has a decimal expansion; for rational numbers show that the decimal expansion repeats eventually, and convert a decimal expansion which repeats eventually into a rational number.

2. Use rational approximations of irrational numbers to compare the size of irrational numbers, locate them approximately on a number line diagram, and estimate the value of expressions (e.g., π2). For example, by truncating the decimal expansion of √2, show that √2 is between 1 and 2, then between 1.4 and 1.5, and explain how to continue on to get better approximations.

*Expressions and Equations – 8.EE*

**Work with radicals and integer exponents.**

1. Know and apply the properties of integer exponents to generate equivalent numerical expressions. For example, 32 × 3–5 = 3–3 = 1/33 = 1/27.
2. Use square root and cube root symbols to represent solutions to equations of the form x2 = p and x3 = p, where p is a positive rational number. Evaluate square roots of small perfect squares and cube roots of small perfect cubes. Know that √2 is irrational.
3. Use numbers expressed in the form of a single digit times an integer power of 10 to estimate very large or very small quantities, and to express how many times as much one is than the other. For example, estimate the population of the United States as 3 × 108 and the population of the world as 7 × 109, and determine that the world population is more than 20 times larger.
4. Perform operations with numbers expressed in scientific notation, including problems where both decimal and scientific notation are used. Use scientific notation and choose units of appropriate size for measurements of very large or very small quantities (e.g., use millimeters per year for seafloor spreading). Interpret scientific notation that has been generated by technology.

**Understand the connections between proportional relationships, lines, and linear equations.**

5. Graph proportional relationships, interpreting the unit rate as the slope of the graph. Compare two different proportional relationships represented in different ways. For example, compare a distance-time graph to a distance-time equation to determine which of two moving objects has greater speed
6. Use similar triangles to explain why the slope m is the same between any two distinct points on a non-vertical line in the coordinate plane; derive the equation y = mx for a line through the origin and the equation y = mx + b for a line intercepting the vertical axis at b.

**Analyze and solve linear equations and pairs of simultaneous linear equations.**

7. Solve linear equations in one variable.
   a. Give examples of linear equations in one variable with one solution, infinitely many solutions, or no solutions. Show which of these possibilities is the case by successively

transforming the given equation into simpler forms, until an equivalent equation of the form x = a, a = a, or a = b results (where a and b are different numbers).

**b.** Solve linear equations with rational number coefficients, including equations whose solutions require expanding expressions using the distributive property and collecting like terms.

**8.** Analyze and solve pairs of simultaneous linear equations.

**a.** Understand that solutions to a system of two linear equations in two variables correspond to points of intersection of their graphs, because points of intersection satisfy both equations simultaneously.

**b.** Solve systems of two linear equations in two variables second pair.

*Functions – 8.F*

### Define, evaluate, and compare functions.

1. Understand that a function is a rule that assigns to each input exactly one output. The graph of a function is the set of ordered pairs consisting of an input and the corresponding output.
2. Compare properties of two functions each represented in a different way (algebraically, graphically, numerically in tables, or by verbal descriptions). For example, given a linear function represented by a table of values and a linear function represented by an algebraic expression, determine which function has the greater rate of change.
3. Interpret the equation $y = mx + b$ as defining a linear function, whose graph is a straight line; give examples of functions that are not linear. For example, the function A = s2 giving the area of a square as a function of its side length is not linear because its graph contains the points (1,1), (2,4) and (3,9), which are not on a straight line.

### Use functions to model relationships between quantities.

4. Construct a function to model a linear relationship between two quantities. Determine the rate of change and initial value of the function from a description of a relationship or from two (x, y) values, including reading these from a table or from a graph. Interpret the rate of change and initial value of a linear function in terms of the situation it models, and in terms of its graph or a table of values.
5. Describe qualitatively the functional relationship between two quantities by analyzing a graph (e.g., where the function is increasing or decreasing, linear or nonlinear). Sketch a graph that exhibits the qualitative features of a function that has been described verbally.

*Geometry – 8.G*

**Understand congruence and similarity using physical models, trans- parenthesis, or geometry software.**

1. Verify experimentally the properties of rotations, reflections, and translations:
    a. Lines are taken to lines, and line segments to line segments of the same length.
    b. Angles are taken to angles of the same measure.
    c. Parallel lines are taken to parallel lines.
2. Understand that a two-dimensional figure is congruent to another if the second can be obtained from the first by a sequence of rotations, reflections, and translations; given two congruent figures, describe a sequence that exhibits the congruence between them.
3. Describe the effect of dilations, translations, rotations, and reflections on two-dimensional figures using coordinates.
4. Understand that a two-dimensional figure is similar to another if the second can be obtained from the first by a sequence of rotations, reflections, translations, and dilations; given two similar two- dimensional figures, describe a sequence that exhibits the similarity between them.
5. Use informal arguments to establish facts about the angle sum and exterior angle of triangles, about the angles created when parallel lines are cut by a transversal, and the angle-angle criterion for similarity of triangles. For example, arrange three copies of the same triangle so that the sum of the three angles appears to form a line, and give an argument in terms of transversals why this is so.

**Understand and apply the Pythagorean Theorem.**

6. Explain a proof of the Pythagorean Theorem and its converse.
7. Apply the Pythagorean Theorem to determine unknown side lengths in right triangles in real-world and mathematical problems in two and three dimensions.
8. Apply the Pythagorean Theorem to find the distance between two points in a coordinate system.

**Solve real-world and mathematical problems involving volume of cylinders, cones, and spheres.**

9. Know the formulas for the volumes of cones, cylinders, and spheres and use them to solve real-world and mathematical problems.

*Statistics and Probability – 8.SP*

**Investigate patterns of association in bivariate data.**

1. Construct and interpret scatter plots for bivariate measurement data to investigate patterns of association between two quantities. Describe patterns such as clustering, outliers, positive or negative association, linear association, and nonlinear association.
2. Know that straight lines are widely used to model relationships between two quantitative variables. For scatter plots that suggest a linear association, informally fit a straight line, and informally assess the model fit by judging the closeness of the data points to the line.
3. Use the equation of a linear model to solve problems in the context of bivariate measurement data, interpreting the slope and intercept. For example, in a linear model for a biology experiment, interpret a slope of 1.5 cm/hr as meaning that an additional hour of sunlight each day is associated with an additional 1.5 cm in mature plant height.
4. Understand that patterns of association can also be seen in bivariate categorical data by displaying frequencies and relative frequencies in a two-way table. Construct and interpret a two-way table summarizing data on two categorical variables collected from the same subjects. Use relative frequencies calculated for rows or columns to describe possible association between the two variables. For example, collect data from students in your class on whether or not they have a curfew on school nights and whether or not they have assigned chores at home. Is there evidence that those who have a curfew also tend to have chores?

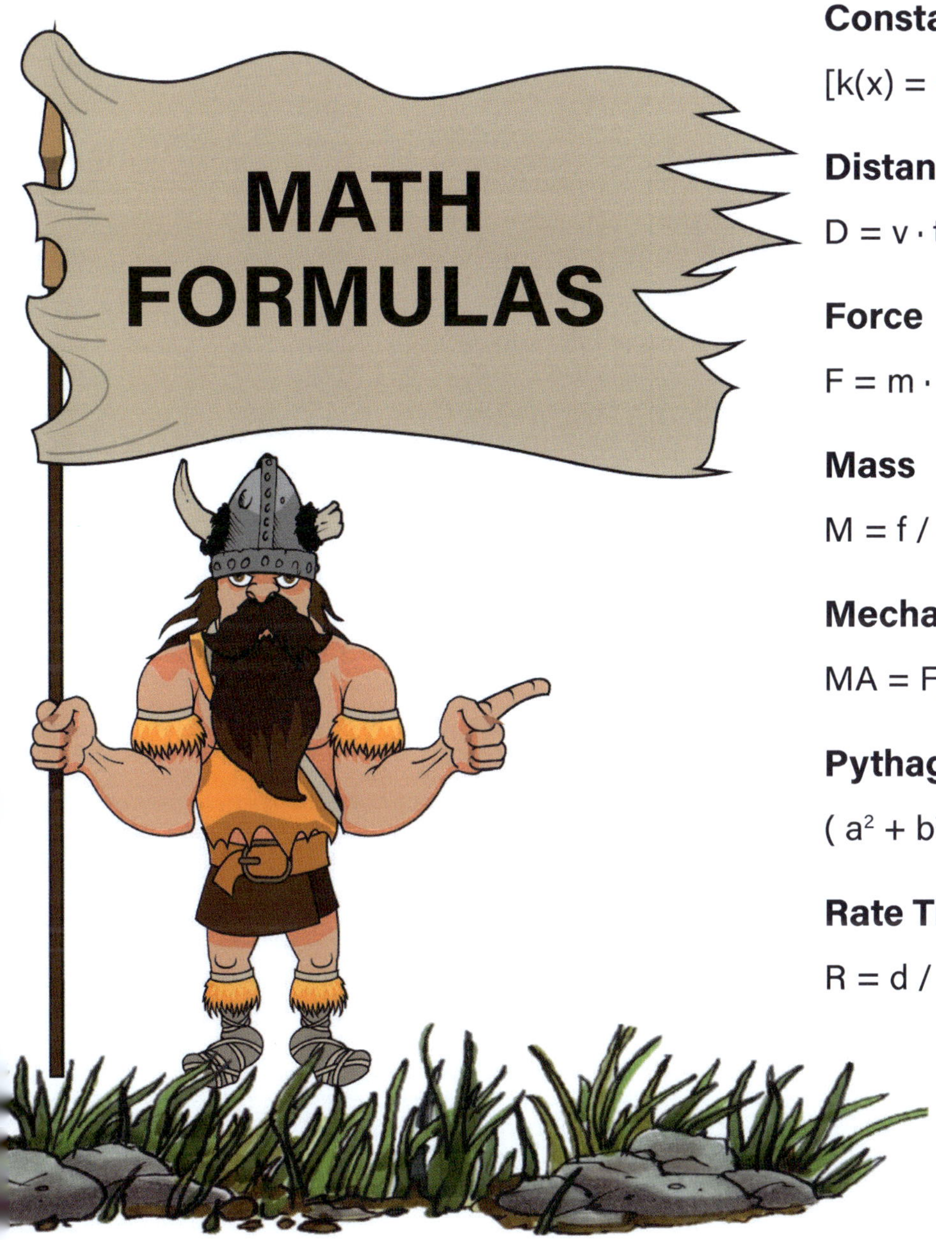


**Acceleration**

$A = V_f - V_i / t$

**Centripetal Acceleration**

$A = v^2 / r$

**Constant of Proportionality**

$[k(x) = y]$

**Distance**

$D = v \cdot t$

**Force**

$F = m \cdot a$

**Mass**

$M = f / a$

**Mechanical Advantage**

$MA = F_r / F_e$

**Pythagorean Theorem**

$( a^2 + b^2 = c^2 )$

**Rate Time Velocity**

$R = d / t \quad T = d / r \quad V = d / t$

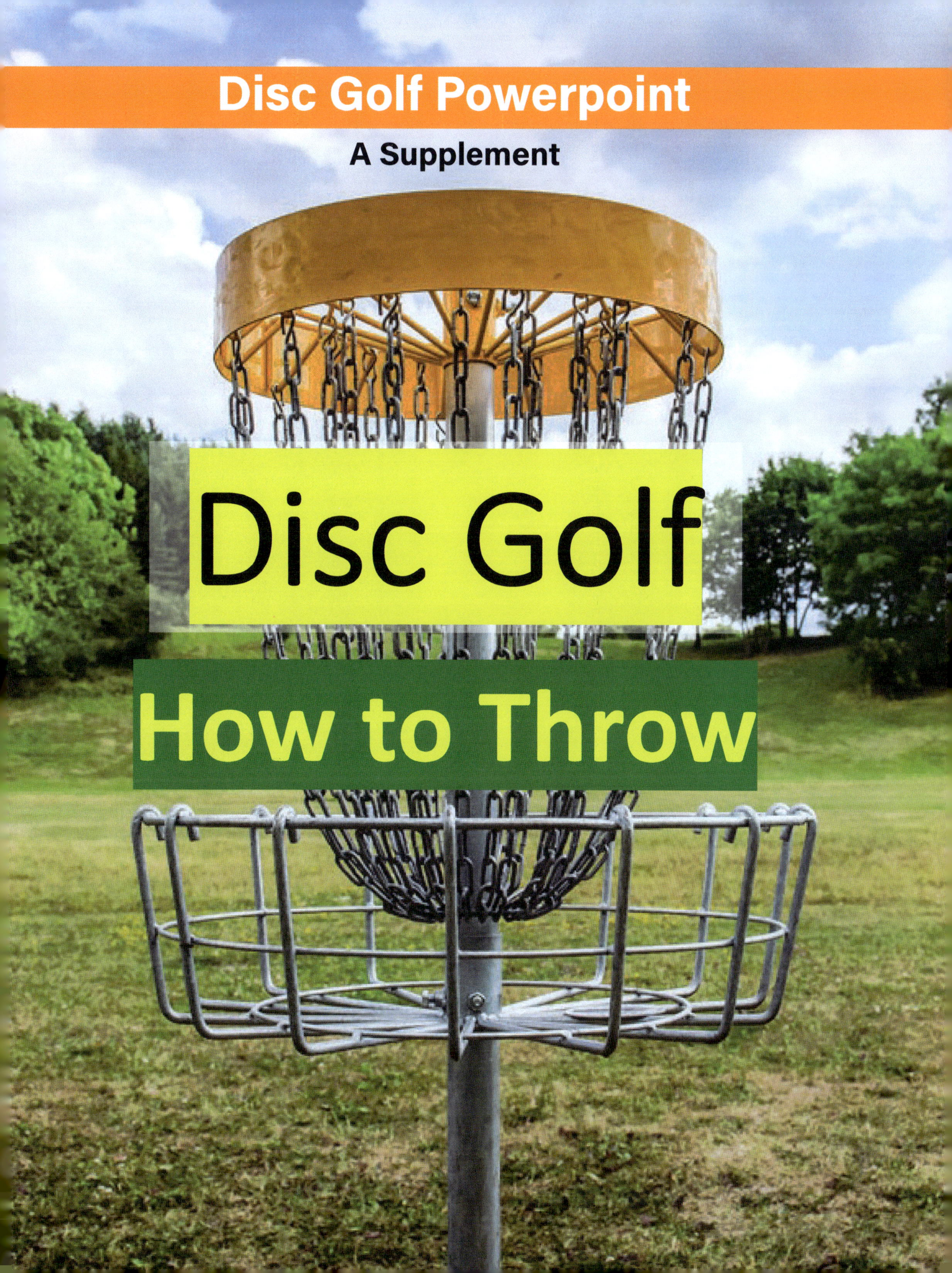
Disc Golf Powerpoint
A Supplement
Disc Golf
How to Throw

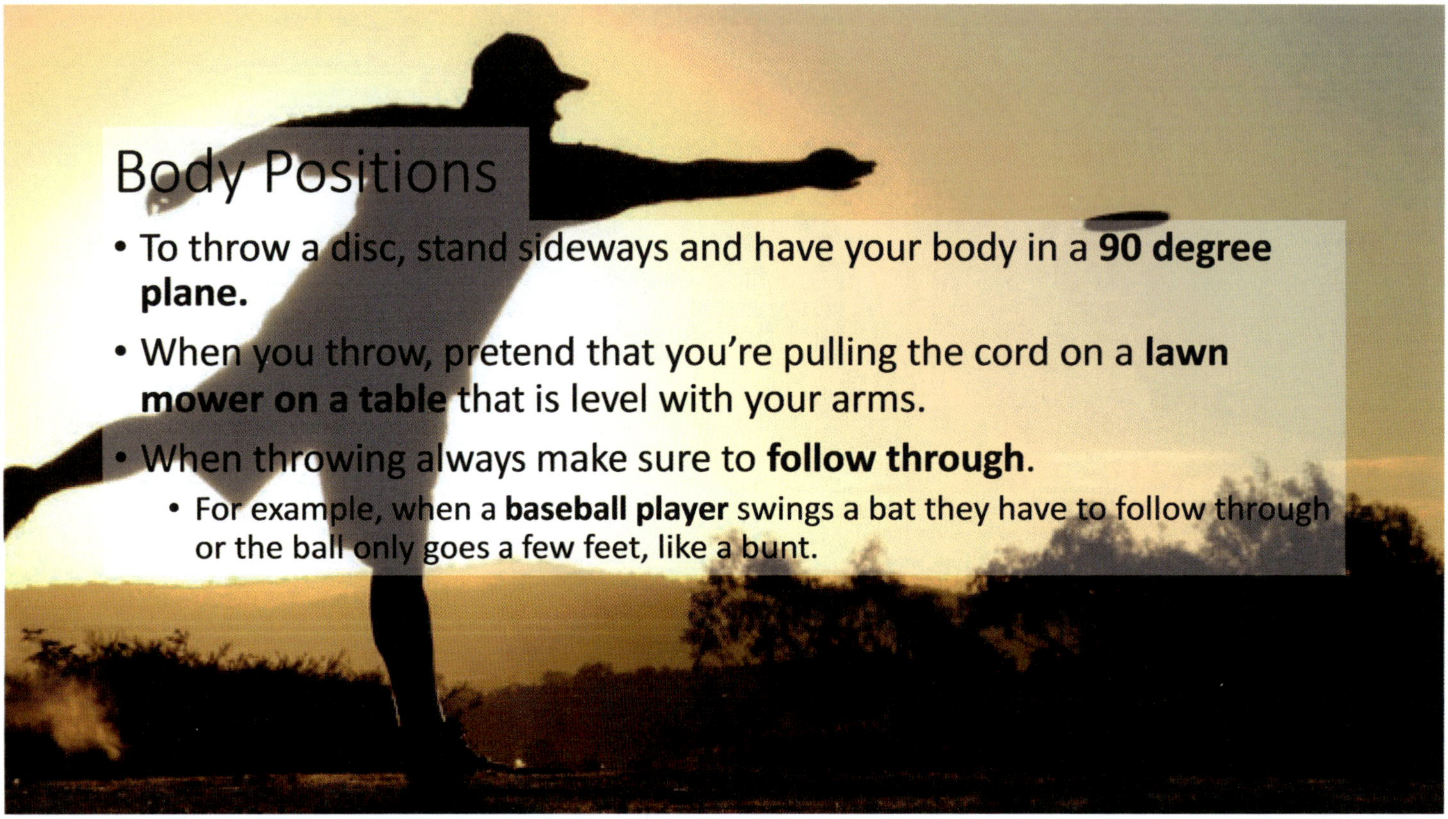
Body Positions
• To throw a disc, stand sideways and have your body in a 90 degree plane.
• When you throw, pretend that you're pulling the cord on a lawn mower on a table that is level with your arms.
• When throwing always make sure to follow through.
• For example, when a baseball player swings a bat they have to follow through or the ball only goes a few feet, like a bunt.

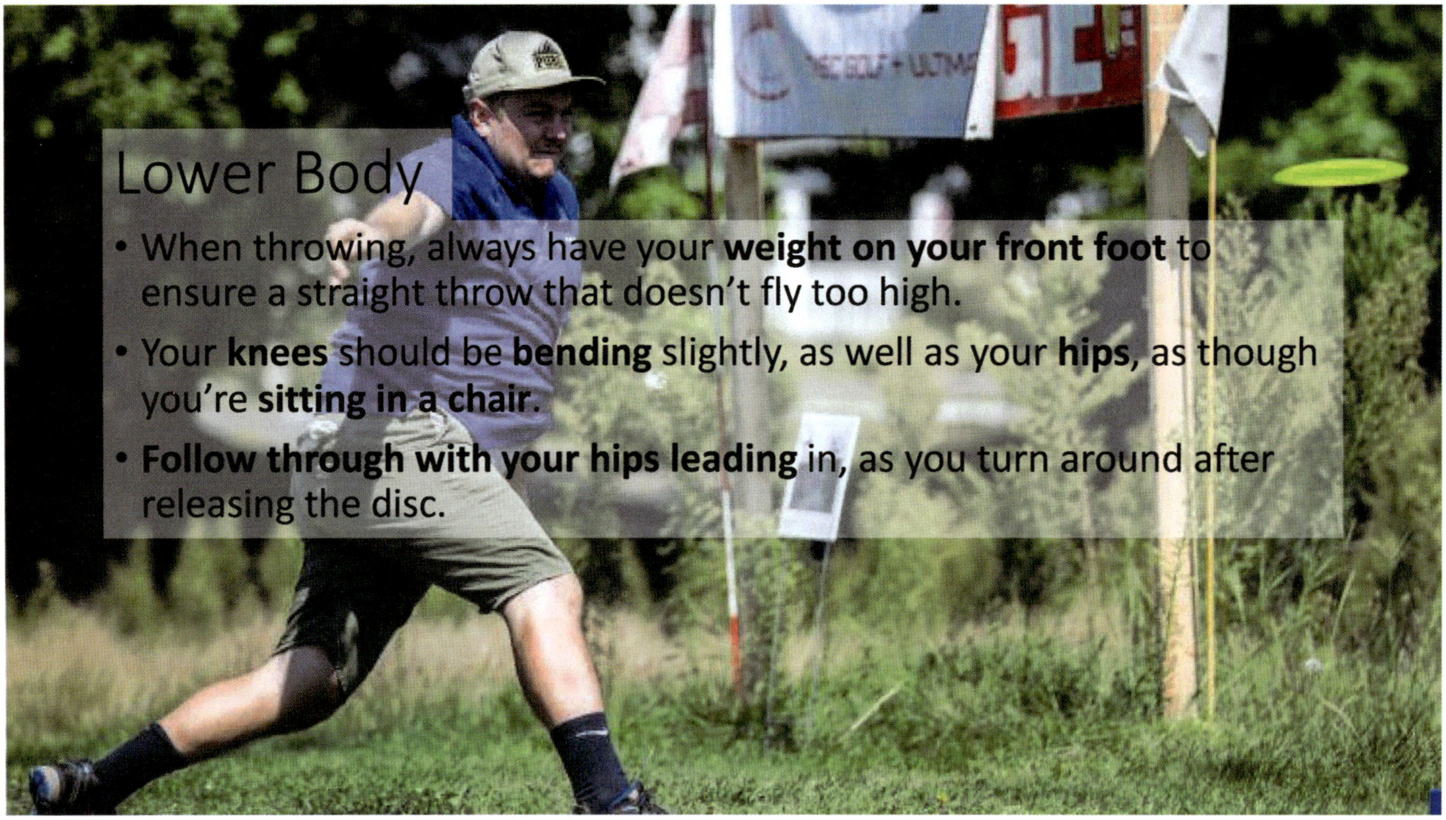
Lower Body
• When throwing, always have your weight on your front foot to ensure a straight throw that doesn't fly too high.
• Your knees should be bending slightly, as well as your hips, as though you're sitting in a chair.
• Follow through with your hips leading in, as you turn around after releasing the disc.

Upper Body
• When you throw, your fore-arm should have a straight platform.
• Your arms should be loose, so the disc glides smoothly through the air.
• When you are throwing, make sure that your trailing arm is pulled back, so your disc doesn't fly straight up.

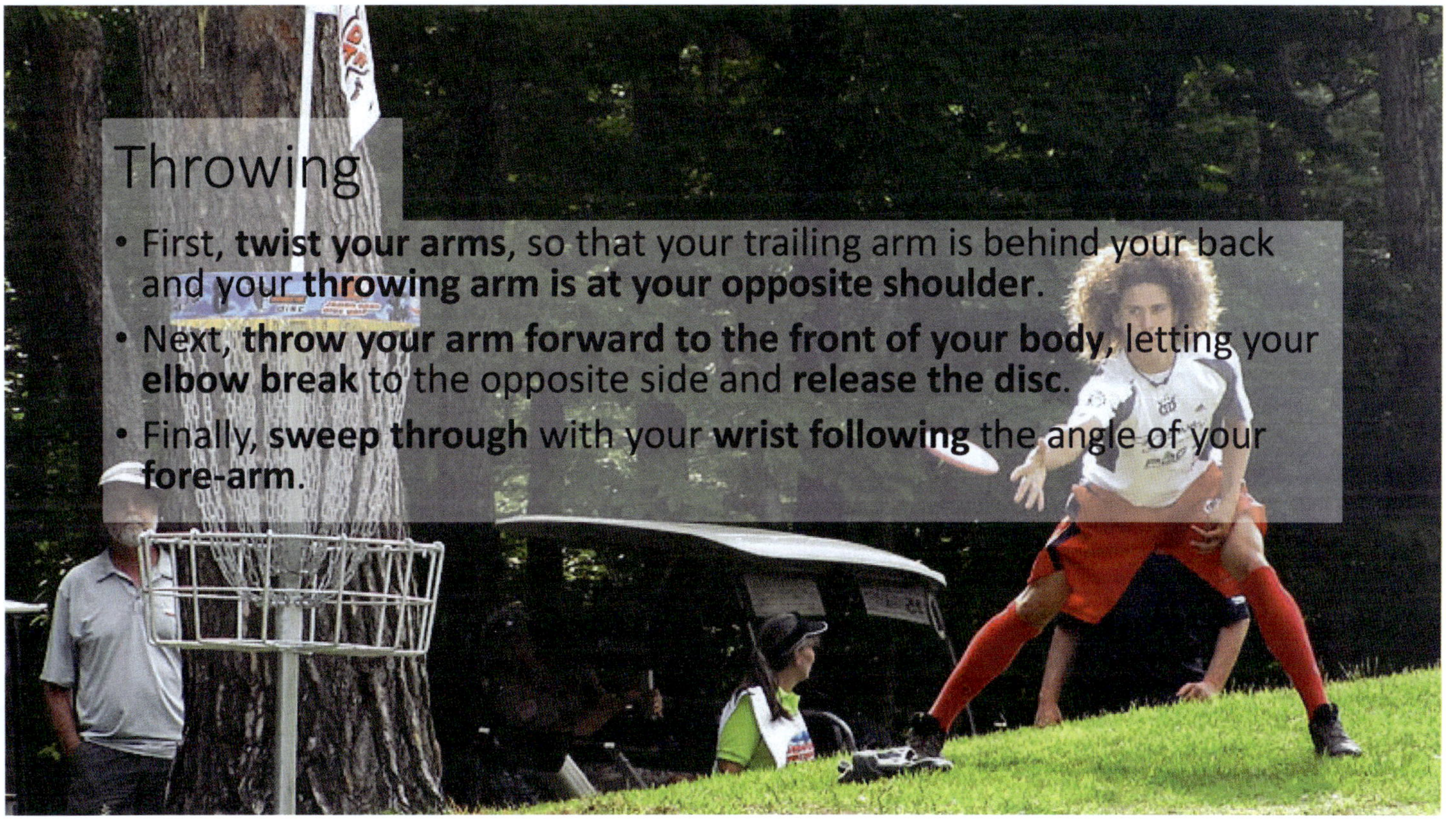
Throwing
• First, twist your arms, so that your trailing arm is behind your back and your throwing arm is at your opposite shoulder.
• Next, throw your arm forward to the front of your body, letting your elbow break to the opposite side and release the disc.
• Finally, sweep through with your wrist following the angle of your fore-arm.

## Mistakes

- **Throwing your arm up too far,** causing the disc to shoot for the stars.
- **Trailing arm isn't behind you** causing a high shoulder angle, which makes the disc go up too high.
- Your **weight is on your back leg,** which drives the disc down towards the ground.
- Your **arm is moving too slowly,** which slows the disc down.

## Easy Fixes

- Keep your arm level, so that your disc doesn't shoot up.
- Keep your trailing arm out of the way to ensure accuracy.
- Before completing your throwing motion, transfer your weight to your front leg as you release.
- Swing your arm quickly to give the disc speed.
  - To ensure quick arm motions try the **towel test**:
    - Tightly hold your towel and swing it like it's a disc (just don't let go!).
    - If there is *one sound* all the way through your swing, you have no power.
    - If there is a *crack* at the last moment, you have power!

Printed in Dunstable, United Kingdom

91218883R00067